Commonsense in Nuclear Energy

Commonsense
in
Nuclear
Energy

Fred Hoyle
Geoffrey Hoyle

W. H. Freeman and Company
San Francisco

Library of Congress Cataloging in Publication Data

Hoyle, Fred, Sir.
  Commonsense in nuclear energy.

  Includes index.
  1. Power resources. 2. Atomic power. I. Hoyle,
Geoffrey, joint author. II. Title.
TJ163.2.H72   1980   621.48   80-11811
ISBN 0-7167-1247-4
ISBN 0-7167-1237-7 (pbk.)

Printed in the United States of America

9 8 7 6 5 4 3 2 1

Published in Great Britain by
Heinemann Educational Books Ltd

Contents

Preface

Historians of the future may well see the discovery of nuclear energy as the most important event of the 20th century, important because it has irrevocably swept away the pre-nuclear era. Before nuclear energy, the portents for humanity had seemed inevitably gloomy. The world population had risen by the end of the 19th century to a level at which energy requirements could no longer be met by what are nowadays called renewable resources, by wood grown in forests, and by wind and water. Society could only continue by making a raid on fossil fuels, first on coal and then on oil. The raid was not just pernicious because it would lead to the eventual exhaustion of fossil fuels. It was savagely pernicious in that it served to drive up the world's population to higher levels still, so that the prospect was one of arriving at fuel exhaustion at the very moment when the population attained by far its greatest number in human history.

In the early years of this century there seemed no way ahead for the human species, no way to the stars, nothing but a disaster of fantastic dimensions, followed by what at best might be the village life of immediate post-medieval times. Nuclear energy changed this apparently hopeless situation within a couple of momentous decades. By 1940, the way to the stars was there for the taking.

It is ironic that, whereas people hardly troubled themselves about the previously-threatened disaster, as soon as the seemingly inevitable prospect of an energy collapse was removed people began to worry. Only an immensely powerful technology could have lifted society almost in an instant out of its previous impasse, and it is the nature of every technology to have negative as well as positive implications. Although nuclear bombs and nuclear energy for civil use both came from the same body of new knowledge, there has never been any question of their practical technologies becoming inadvertently mixed. At a political level, however, there is an important way in which the bad and the good really can become intertwined. If the developed nations really do run into a desperate energy shortage, the tensions arising from the lop-sided geographical distribution of dwindling fossil fuels are only too likely to lead to presently-existing bombs being launched on the world. Energy shortage seems a sure prescription for nuclear war.

Cumbria, 1979
Fred Hoyle
Geoffrey Hoyle

I

The ultimate price of an energy collapse

It is somewhat natural to think that the only energy we use is that which we control ourselves, as for instance when we switch on some electrical gadget or take a ride in the family car. Most of the energy we need is hidden from our immediate view, however. It goes in digging metallic ores out of the ground, smelting metals, growing food, industrial manufacturing, the transport and distribution of goods, and so forth. These other forms of energy-use comprise a good 60 per cent of our needs, and these other ways cannot be subject to any simple conservation of the kind: 'I will only switch on the electric light when I really need it.' A big effort towards personal conservation by each of us might make a 10 per cent difference to the total energy requirement of the whole community. It is an error therefore to think that such an effort could obviate the need for the so-called 'hard' technologies of oil, coal, and nuclear.

An all-out conservation effort on a national scale could produce a sharp fall of a few per cent in energy demand, which in a commercial world of oil prices highly sensitive to demand would be an effective short-term political manoeuvre. There is no way, however, in which conservation could be a long-term answer to our energy problems.

The policies which the anti-nuclear movement is seeking to impose on society will, if they are permitted to succeed, lead to energy shortages in the 1980s, to energy crises in the 1990s, and to energy disaster in the early 21st century. The world population has increased over the past few hundred years from some 500 million to a present-day total of about 4000 million. The population has been able to rise quickly because of the availability of coal and oil. In the year 1600 there might have been some argument for restricting energy supplies to prevent the world's population explosion. Today, such a crude method of restricting population growth is too late. The explosion has happened already, and if energy for agricultural machinery and for the machinery of industry were suddenly taken away from us now, most of the present world population would have to die, not over the natural span of three score years and ten, but quickly.

People in the north-east of the United States experienced a temporary energy shortage during the very cold winter of 1977–78, and owing to a short-lived strike by tanker drivers the British were inconvenienced by a lack of petrol and heating oil in early January 1979. It is to be emphasised that these passing phases gave only the very faintest indications of what a true energy shortage would be like.

In an earlier book* one of us remarked that energy is more important than money. The truth of this precept has recently been brought home to the financial world by the steep decline of the U.S. dollar, a decline caused in the main by insistent demands for increasing quantities of oil, imported by the U.S. from the Middle East, Africa and S. America. Once again, these demands do not come from a true energy shortage, let alone an energy crisis, because the U.S. has reserves of both coal and nuclear fuel. The storm in the money markets has come merely from the *kind* of energy at issue—the demand is for oil, not for lumps of coal. A genuine shortage of all forms of energy would be an incomparably more serious matter.

The day will certainly arrive when there is no more coal and oil to be used by the machines upon which our society has become dependent. How soon that day will come turns on social issues that are still unresolved. At present, four-fifths or more of the world population is poor, with annual per capita incomes in the subsistence range from $100 to $500. If this big majority remains poor, oil reserves will last for the benefit of the one-fifth that is rich for some 30 to 50 years, and the world's total recoverable coal resources will last for about 200 years.† If on the other hand the whole world population attains to what in the developed countries is regarded as a reasonable standard of living, the much larger energy consumption rate then needed would lead to the exhaustion of all reserves, coal and oil alike, within a few decades, by about the year 2025.

There is much present discussion of what new energy form will eventually replace coal and oil. Will it be solar? Or wind and water—solar also in disguise? Or nuclear fusion? Or nuclear fission? Of these possibilities some are clearly inadequate, and only nuclear fission is definitely known to be viable. In the course of this book we hope the reader will decide who it is that is behaving responsibly. Is it the governments of W. Europe who are aiming to develop nuclear fission, and the American utility companies that would also like to develop nuclear fission? Or is it the anti-nuclear activists, abetted by some politicians and by certain sections of the media, who would stop if they could the only proven source of energy for the future?

The crucial point is not to eke out oil supplies for a few more years, or coal even for a century. The crucial point is that inevitably down the years the snows of winter will continue to fall and bitter winds will continue to blow long after the coal and oil is all gone.

---

* F. Hoyle, *Energy or Extinction*, Heinemann, London, 1977.
† For a recent estimate of recoverable resources of coal, see E. D. Griffith and A. W. Clarke, *Scientific American*, Volume 240, page 28 (1979).

2

Desecration of the environment

The view is widely encouraged by the anti-nuclear movement that everyone sensitive to the environment should be opposed to the development of nuclear energy. By twisting this dogma around, the inference is then made that supporters of nuclear energy are hostile to the environment. We propose to show here that this inference by inversion is false.

Every aspect of technology is in some degree a blot on the environment. Because we are all dependent on technology for our very existence, there is no prospect of removing all pollution. Consistent with the technology that we must have willy-nilly, the problem therefore is to minimise pollution of all kinds. So far as the visual aspects of the environment are concerned, of all presently-known sources of energy, nuclear fission is by a large margin the least obnoxious.

Of the several million people who annually visit the Lake District National Park in the U.K., we doubt that one in a thousand is visually aware of the nearby nuclear establishments at Windscale, the largest of their kind in Britain. Yet a considerable fraction of visitors soon becomes uncomfortably aware of the desecration of the Park caused by the City of Manchester. Of all places in the Lake District, Thirlmere is perhaps the most depressing and gloomy. Natural lakes do not have the sterile white band around their edges to be seen in Figure 2.1. Plants at the edges of natural lakes accustom themselves to seasonal variations of the water level, but plants cannot accustom themselves to the sporadic variations imposed on Thirlmere by the demands of the City of Manchester.

Thirlmere is a water-storage lake, not an energy-supply lake, but a similar sad state of affairs can be found in scores of valleys in the Highlands of Scotland, the same sterile white bands along the sides of the lochs, the same unsightly dams used to raise water levels, thereby inundating the beautiful lochside paths trodden by former generations of Highlanders. Energy supply, hydroelectric energy, was the cause of this visual pollution. Yet if all the hydro-power output from all these desecrated valleys is added together, the total does not amount to more than the output from a single nuclear station. The situation existing in the

Figure 2.1   Thirlmere, showing the desecration of a once-beautiful lake. Natural lakes like Ullswater have vegetation down to the water line.

Highlands is a prime example of how to *maximise* visual pollution. So far as we are aware, there have been no storms of protest about it.

Nor were there any protests, any marching in British streets, to protest Parliament's recent decision to vote money for a design study to be made of a project to dam the mouth of the River Severn, a project that would also be large in its adverse visual impact on the environment and yet very small compared with our energy requirement. By a like token there will need to be many hundred miles of unsightly structures placed around the British coast, if energy from oceanic waves is ever to amount to much. True, it seems to be the intention to begin these constructions in the region of the Hebrides, where they will not be seen by most people, leaving the Highlanders once again with the resulting environmental disaster. For anyone professing concern for the environment to support projects like these is indeed anomalous.

Turning from visual pollution to other forms of environmental desecration, in the majority of published recommendations for wind, water and solar energy, the point is made that such energy systems put no excess heat into the environment, with the implication that nuclear energy on the other hand pollutes the environment by heating it. If the world were to derive all its present energy requirements from fission, the input of energy to the environment would raise the average temperature of the Earth by less than 0.01 °C, a wholly trivial effect. Local heating close to a reactor would be greater, however, and it is worth considering how much local heating there might be.

It is considered desirable in Britain to site nuclear reactors near the sea, in order to have ample water available to take excess heat away. For the future, however, one can contemplate that excess heat might be piped to homes in need of it, and this would be at least as practicable a scheme as most non-nuclear ideas. But let us suppose all reactors to be on the coast, disposing of their excess heat into the sea. If Britain derived its energy entirely from nuclear fission, about 200 reactors would be needed. If spaced uniformly apart along the coastline of the U.K., which has a length for this purpose of about 5000 miles,* the reactors would be separated from one another by 25 miles, although they could of course be placed in clusters if it was thought desirable to do so, with correspondingly greater distances between the clusters. If the excess heat from the reactors were fed uniformly into a strip of water 5 miles wide everywhere along the coastline, the temperature offshore would rise by about 2 °C. Without taking trouble to spread the heat uniformly through such a strip, however, there would be warmer pools of seawater around the site of each reactor. In these pools the temperature might rise locally by as much as 10 °C.

Environmentalists object to the generation of warm pools, arguing that the local ecology at the sites of reactors would be destroyed. More accurately, the local ecology would be changed. Fish preferring the present cooler waters would move away from the warm pools, while sea creatures requiring higher temperatures than those which presently exist around the British Isles would come to establish themselves in what for them were favourable localities. The ecology of the coastline would therefore become more varied, and some people might consider the greater diversity an advantage rather than a disadvantage. Moreover, once it was understood that there was no danger in bathing in the warm pools, people who now rush south to the Mediterranean for their summer holidays might also consider the new situation to be an advantage.

To be intensely worried about cold-water fish moving a few miles up or down the coast, which movements the fish are making all the time anyway, must require a remarkable measure of sensitivity. So much so that one would be perpetually harrowed with concern for the plight of other animals in much worse circumstances. One might well be led for instance to argue the need for injecting excess heat into the environment in order to relieve the very real plight of birds in bitterly cold weather.†

The cull of grey seals which the British government recently proposed to carry out off the north coast of Scotland was stopped through the intervention of *Greenpeace*, an environmentalist movement. This intervention has been criticised on ecological grounds by C. Summers in the issue of *New Scientist* for 30 November, 1978. Dr. Summers points out

---

* The length of the coastline depends on how far one counts minor inlets. Such indentations are not relevant here, and are not included in the estimate.
† All our wrens perished in the winter of 1978-9.

that, without the cull, the population of grey seals will tend to rise out of bounds, eventually depleting severely the stock of fish upon which they feed, and eventually as the fish decline running the seals into starvation. Be this as it may, it is hard to see how *Greenpeace* can claim an intense sensitivity to the welfare of fish, to the neurotic degree that it worries them if fish are required to move a few miles up or down the coast, and at the same time for them also to claim concern for the well-being of the grey seal which kills the much-loved fish in considerable quantity. We suspect a deal of humbug is being talked.

For the town-dweller, having little contact with animals, except possibly for a pet dog or cat, it is easy to take a relaxed, sentimental position. For fishermen and farmers the situation is different. Without continuous protection, every domestic duck and goose in the Lake District National Park would within an hour or two and in broad daylight fall prey to foxes. The Lakeland farmer cannot therefore take a sentimental position over foxes. Nor can the gardener permit greenfly, wasps, and other pests to increase beyond bounds.

Some issues concerning animals are fairly clear-cut, while others are not. For the clear-cut cases it is reasonable to take clear-cut decisions. For as long as the Japanese hunt the whale to extinction the authors of this book will buy no Japanese goods, a position we hope other environmentalists will also take. And the refusal of the French to sign an international agreement prohibiting the shooting of migratory birds caused us to desist, at any rate temporarily, from drinking French wines.

Other cases are much less certain. A few years ago it was thought that herds of cattle throughout England had been freed from bovine tuberculosis. Then an outbreak of the disease occurred in the West, raising the urgent question of from where the bacillus responsible for the disease had come. By chance a badger killed in a road accident was found to have been suffering from bovine tuberculosis. Further investigation then showed that some other badgers in the West of England were similarly infected, but not badgers in the rest of the country. It seemed a fair inference that badgers had provided a reservoir for the disease, and that after the initial eradication of it, the cattle had been subsequently reinfected through grazing on land over which badgers had passed. On this reasoning, the department of the British government concerned with the cattle ordered the extermination by cyanide-gassing of all badgers throughout the West of England.

Likely enough, the government reasoning will be found correct, although it is an elementary point of logic that both cattle and badgers might have contracted the disease from some third source. Before starting the gassing, this possibility should have been eliminated, we would think. Perhaps it was, but if so the refutation should have been clearly published.*

We have not seen *Greenpeace* assembling its forces to prevent the

---

* *Note in proof:* This criticism has been recognised by the present government, which has formed a committee to report on the matter.

gassing of these wretched animals. We think the leaders of this move-ment are astute enough to realise that extensive media coverage would be lacking. Likely enough, most people would sooner have badgers gassed than risk contracting bovine TB themselves, or risk their children and relatives contracting it. Recently there has been a sharp increase of human tuberculosis, also in the West of England. Since human tuberculosis does not come from badgers, the *prima facie* case that badgers are not the cause of the disease in cattle may be considered to have been strengthened. The grey seals were a better case for *Greenpeace*, because grey seals are conveniently removed from our everyday lives, if not from the lives of the fish over which the anti-nuclear movement claims to be so acutely concerned.

By far the greatest environmental pollution comes from vehicular traffic, from the roar and stink of it through the centre of every town and city in the land, and from the 7000 deaths and 350 000 injuries which it causes in Britain each year. Unlike the production of energy, which is essential to the continuation of civilised society, most city-centre traffic is mere convenience-traffic; that convenience-traffic should so dominate our existence is the scandal of the age. We would take the tough environmental position that, with the exception of buses, taxis, ambu-lances and medical traffic generally, police cars, and delivery vans and lorries at certain specified hours, traffic should be banned entirely from city centres.

This then is the environmental position taken by the authors, the background against which we now go on to develop our views on nuclear energy.

3

Nuclear reactors cannot explode—but other things can

In *Energy or Extinction* one of us remarked that a nuclear reactor is no more likely to explode than a bar of chocolate. The point has been further developed by Lord Rothschild in his recent Richard Dimbleby Lecture.* To quote:

'Let me give you an example of an utterance by what I will charitably call an emotionally over-committed man:

WE ALMOST LOST DETROIT
By John G. Fuller, Reader's Digest Press, $8.95

and, at the same time, the follow-up by an emotionally over-committed lady, in the press; in this case the *New York Times* book review, on 30 November 1975. There had been an accident in the small Fermi nuclear reactor, near Detroit. No one was injured, nor were there any serious consequences outside the reactor. The reviewer, Mary Ellen Gale, said:

> When things went awry at the Enrico Fermi reactor near Detroit, four million people went about their business in happy ignorance, while technicians gingerly tinkered with the renegade's invisible interior. They knew what the public did not—a mistake could trigger a nuclear explosion.

In fact, a nuclear explosion is no more feasible in a nuclear reactor than it is from chewing pickled cucumber or gum.'

At the level of Lord Rothschild's lecture, the opponents of nuclear energy would admit that nuclear reactors do not explode,† but at the level of a *New York Times* book review, and rather generally in the media, the notion that they might is encouraged.

The recent accident to Metropolitan Edison's Unit 2 at Three Mile Island near Harrisburg, Pennsylvania (March 28, 1979) provided the media with the opportunity to disturb the public in a deep and serious way. There were indeed serious aspects to this accident which we shall

---

* *The Listener*, 30 November 1978, page 717.
† For example, Sir Martin Ryle, *The Daily Telegraph*, 28 December 1978.

discuss in Chapter 7, but they were technical in their origin and effects. There was no death list such as occurs almost daily in real disasters, and in Harrisburg the much publicised release of radioactivity into the environment was small. According to a statement from the U.S. National Academy of Engineering (April 12, 1979) radiation doses experienced by the public were such as would be received quickly from the natural background. Put in another way, spending a week or two's holiday in regions of abnormally high radioactive background, such as Dartmoor, certain parts of Cornwall, and the Isles of Scilly, would expose the reader to excess doses of radiation that were greater than the people of Harrisburg itself received from the accident at Three Mile Island.

The total misconception that nuclear reactors can explode is at the heart of popular hostility to nuclear energy. Environmental nit-picking, like pools of warm sea-water around the sites of nuclear reactors, are good enough issues for tub-thumping (Figure 3.1), but the real problem is that the dark fear has been established already in the minds of people that a reactor is really a nuclear bomb in disguise.

Although nuclear physics is a comparatively new branch of science, it is in many ways simpler than the older sciences of chemistry and biology. Unlike chemistry, where very many combinations of substances are explosive, there is only one basic way to produce a nuclear explosion, a way that is most carefully avoided by the designers of nuclear reactors. This is why one can be clear that reactors cannot explode, just as one can be clear that an unaided human cannot jump a thousand feet up into the air.

In our daily lives we handle chemicals in ways that border all the time on disaster, and we do so quite deliberately. Passengers in an aeroplane

Figure 3.1 Tub-thumping.

know full well (or if they don't they should) that great quantities of explosive liquid would be released from the fuel tanks should the plane be forced to crash land, or if the plane were hit in mid-air by even a comparatively small object. This happened in early October 1978 over Lindbergh Field, San Diego, California. The issue of *Time* for October 9 had the following description:

DEATH OVER SAN DIEGO

A freak collision kills 150—even when everybody is following the rules
   'Tower, we're going down. This is PSA.'
   That terse message from Pacific Southwest Airlines Captain James McFeron was delivered in the flat, cool tones cultivated by professional pilots. It conveyed no more emotion than McFeron had expressed a few moments before in asking for clearance to land. Yet now his 66-ton Boeing 727 jetliner with 135 'souls on board', according to the jargon of the aviation industry, was hurtling out of control at 280 m.p.h. towards San Diego's residential North Park neighbourhood. It was already on fire.
   'We'll call equipment for you,' replied the tower controller at Lindbergh Field in that same business-as-usual manner. The final word from Pilot McFeron: 'Roger.'
   Seconds later, an air traffic control specialist at the airport peered into his radarscope and got his first glimpse of what was happening. As his screen displayed the falling and fragmenting wreckage of two aircraft that had collided at 2,650 ft. three miles northeast of Lindbergh Field, he muttered, 'Jesus Christ, an aluminium shower.'
   The hellish orange flames and oily black smoke that rose quickly into San Diego's sunny but smoggy skies one morning last week signalled one of the worst air tragedies in U.S. aviation history. At least 150 people died, the first fatalities on PSA's record. They included all 135 aboard the PSA airliner, the two occupants of a tiny 2,100 lb. Cessna 172 that had collided with it, and at least 13 residents struck by aircraft debris or engulfed by the flames that destroyed ten houses.

·   ·   ·   ·   ·

At the subsequent enquiry in early December 1978, an official from the U.S. government department responsible for airport safety was asked why a sophisticated modern piece of radar equipment which might have prevented the accident had not been provided for Lindbergh Field. Astonishingly, the reply was that there were so many airports where such an accident could equally well have happened that to fit them all with sophisticated radar would go far beyond the financial resources of the department. There was no marching in the streets of Washington to protest about this situation. Nor was there an outburst from the media such as that which followed the radioactive leak at Harrisburg on March 28, 1979.

It is difficult, even for a professional chemist, to appreciate just how wide are the possibilities for chemical explosions. The explosion at Flixborough, S. Humberside, occurred on Saturday night, June 1, 1974, and on Monday, 3 June, the London *Times* published the following report:

'Senior British and Dutch executives associated with one of the world's worst disasters in the chemical industry last night said they were totally baffled at the cause of the explosion in a plant they believed to be fail-safe, but in which 29 people died, more than 40 were injured, and hundreds of acres of surrounding land were devastated.

'Last night, 24 hours after the explosion, a smoke pall visible 20 miles away and bearing a marked resemblance to pictures of an atomic explosion cloud, was drifting across the Lincolnshire plain under a strong south-west wind, and the crews of more than 40 fire engines were still fighting flames from burning chemicals.

'A radioactive fallout scare spread through the villages adjoining the plant yesterday. Last night at the news conference executives said there was a container of radioactive material in the plant, but that it had been found intact and safe.

'Nypro (which operates Flixboro') is a specialist company jointly operated by the National Coal Board and Dutch State Mines to produce a material called caprolactam, which is essential for the production of man-made fibres using a material called cyclohexane, which in turn is a by-product of coal-coking processes and is similar in its general properties to petrol.

'The seat of the explosion was traced yesterday to a part of the 20-acre factory where the cyclohexane is oxydized by being heated and subjected to pressure. There, in section 8 of the factory, it is believed 29 victims died instantly, their bodies probably disintegrating in the intense heat. Forty others escaped from neighbouring departments, most of them being burnt, injured or badly shocked. Ten were still in hospital last night. The intensity of the fire late last night was still preventing any access to the area where most of the victims died.

'Had the explosion happened on a weekday, the death and injury toll would have been much higher, as the plant employs a total of 300 people, with about 200 on duty on a normal day shift. On Saturday night there were only 70 on the site.

'Mr. Wym Bogers, Chairman of the Dutch State Mines, said last night: "It is completely amazing. We cannot understand it".'

. . . . . .

If the chemical industry had traditionally employed a fraction of the care which the nuclear industry has taken from its beginning, the cause of the Flixborough disaster would not have seemed so amazing. So far from the explosion having a nuclear cause, the one thing which came intact and safe through the holocaust was a container of radioactive material.

Most chemical plants exude explosive gases, usually only in small amounts. If the gases are lighter than air, they soon rise upward and become widely and harmlessly distributed. Heavy explosive gases tend to fall downward, however. In such cases it was thought that, provided the heavy explosive gases were always exuded out of doors, never in a closed, confined situation, natural air movements would also disperse them in a harmless way. For the most part this supposition was correct. Under conditions of exceptionally still air, however, a pool of heavy explosive gas could accumulate close to ground level. Possibly such pools had accumulated at Flixborough many times before without being detonated until the night of June 1, 1974.

The reader should not be misled by high-sounding names like caprolactam and cyclohexane, or by reference to a 'specialist company'. The plain fact is that Flixborough was making the base material for the production of nylon. As one commentator remarked at the time, 'Flixborough is the price we have to pay for women's stockings.'

There are many examples of lethal chemical explosions not involving specialist chemical companies, recently of an ice-cream van and of a whisky store. On December 22, 1977 an explosion killed 23 people in a New Orleans grain elevator. The book

WE ALMOST LOST NEW ORLEANS
by John G. Fuller, Reader's Digest Press, $8.95

has not yet appeared. When it does, it will doubtless be reviewed by Mary Ellen Gale.

On July 11, 1977, a tanker lorry carrying liquid gas exploded in the centre of a crowded camping site at San Carlos de la Rápita, E. Spain. By July 14 the death roll had risen to 140. The issue of *Time* magazine for July 24 carried the following report:

IT WAS LIKE NAPALM
It was approaching mid-afternoon, and a sparkling, early-summer Spanish sun still shone high over the tiny Mediterranean resort of San Carlos de la Rápita. Most of the 600 French, West German and Belgian tourists at Los Alfaques (the Sandbars) campsite were eating a leisurely sitdown lunch in front of their tents and trailers or at picnic tables under the shade of palm and cypress trees. Others were dozing off for a vacation siesta. Groups of children romped among the sunbathers basking on a narrow beach.

At exactly 2:36 p.m., a 38-ton tanker truck carrying 1,518 cu. ft. of highly combustible propylene gas from nearby Tarragona to an industrial refinery in central Spain peeled around the long bend of the highway behind the camp at 40 m.p.h. and skidded out of control. Perhaps already on fire, it crashed into a retaining wall, rolled and, as it exploded, spewed torrential fountains of fire that washed across most of Los Alfaques. Flames towering hundreds of feet engulfed vacationers and their gear, setting off a secondary round of blasts from exploding butane cookers and automobile gas tanks. Parts of the tanker were blown almost half a mile away. Trailers were burnt to their frames in an instant, like paper models. Campers ran into the water to douse flames on their bodies, only to be burned even more severely by the chemical reaction.

'It was like napalm, it was an inferno,' said a French visitor who had been washing dishes in a trailer that was spared at the edge of the camp. 'People were running everywhere, screaming, some of them on fire.' More than 100 were killed on the spot, most burnt beyond recognition. Another 150 or more lay writhing in the havoc, grotesquely scorched. In all, the fire storm that devastated Los Alfaques had killed 144 by the week's end, and left some 75 injured, many critically. Not since a pair of jumbo jets collided and caught fire on a runway on the Spanish Canary Island of Tenerife in March 1977, killing 582, had there been a burn disaster of such proportions.

· · · · ·

What needs especial emphasis about this 'burn disaster' is not that it happened, but that its happening was entirely predictable. Not predictable of course at San Carlos de la Rápita, but predictable somewhere. So long as compressed gas is moved freely from place to place, it is inevitable that such incidents will occur. With a change of venue and a change of circumstantial details, an essentially identical accident had occurred some months earlier in the United States. In the issue of *Time* for March 13, 1978 there appeared the report:

PLAYING RAILROAD ROULETTE
While workers were preparing to transfer 20 000 gal. of liquid propane gas from a derailed tank car to trucks in Waverly, Tenn., the gas suddenly exploded levelling 14 buildings. 'It was just like you were thrown into a furnace,' said Truck Driver Carl Stokes, who was burned severely. 'It was like a power throwing us into the sun. People were walking, their clothes were gone and their bodies were completely burned.' The toll: twelve dead and at least 50 injured.

· · · · ·

People have sat down in their thousands to prevent the construction of a nuclear reactor at Seabrook, New Hampshire, and this they have done on a *zero* total for victims of nuclear accidents. No-one has sat down to prevent tankers with liquid gas from using the roads, even though the death roll for liquid gas runs to many hundreds—'it was like a power throwing us into the sun'.

The U.S. Judiciary halted construction of the Seabrook reactor at one point, in a decision which ex-Governor M. Thompson of New Hampshire described as 'asinine'. The same judiciary has not sought to interfere with the transport of gas, however. If in the cold of winter it tried to do so, some of its members would no doubt be quickly out of a job, with beneficial consequences for the United States.

4
Death sentences

The underground mining of uranium, the source of nuclear energy, is less than one-tenth as dangerous as mining coal, and strip mining for uranium is less than one-hundredth part as dangerous as coal mining underground. These statements refer to relative risks calculated for equal output not in tonnage but in energy; nuclear energy equals coal energy. The relative risks are also calculated for the poor efficiency of uranium utilisation discussed in Chapter 14. For the much greater efficiency given by breeder reactors, the relative risk of uranium mining would be so small as to be negligible.

In the past about 140 coal miners were killed each year in the United States. Although this number is much less than it used to be and is currently being reduced, there will still be many unnecessary deaths of American miners for each decade that coal is preferred to uranium as an energy source. This is the death warrant which the Carter Administration issued when it decided to develop coal as the primary source of energy in the U.S.

The dangers of coal mining have not been stated better than they were in *Harper's Weekly* just after the worst mining disaster in American history, which occurred at Monongah, West Virginia. In their book *Disaster Illustrated*,* W. Welman and B. Jackson set the scene at Monongah:

'All the local people knew about the hollow hills. The Consolidated Coal Company of Boston had honeycombed them with mine shafts until it seemed the town might one day fall in. Even so, the 6,000 people who lived there were all connected with mining. Every family had at least one man who worked below ground.

'Early on Friday, December 6, 1907, 380 men entered the No. 6 and No. 8 mines of Consolidated. The two mines were connected underground, but one had its entrance on one side of the South Fork River, the other on the opposite side. Everything seemed to go as usual until 10 a.m. when an enormous explosion rocked Monongah. Everyone knew at once what it was.

---

* Harmony Books, New York, 1976, p. 174.

'The townspeople ran for the mines, and there they saw the props holding up the entrance to No. 6 had been blown clear across the river. The entrance to the other mine (No. 8) was likewise a shambles, and both were blocked by tons of earth, rock and coal. In the hills, however, there were several openings to No. 8. Such fervid hope was born in the hundreds of hysterical kinsmen who rushed to these holes that they could almost see the trapped men pouring out of the openings. But all that poured out of the entryways was volumes of the 'black damp', the poisonous methane gas that had probably started the coal-dust explosion. At every opening, withered vegetation showed that the gas was there. No man could make his way through those fumes.

'A huge rescue force gathered and set to work at every point of entry and exit. Almost at once, six bodies and five injured men were pulled to the surface. The injured men could remember little of what had happened, and merely spoke of crowds of struggling frantic men below them. It was guessed that they were not very deep in the mine when it exploded.

'In No. 6, rescuers discovered that the explosion had knocked out the ventilating system, and no one could stay long in the mine, which was now filled with bad air. The rescue workers broke up into teams, and went in short relays into the sickening air below. Each time, the groups tried to penetrate a little deeper, and each time they returned without a survivor.

'The efforts of the rescuers were heroic, but mostly in vain. Only 19 miners survived, most of whom had escaped immediately after the explosion. The bodies in the mine—361 of them—were torn and mangled and blackened. The majority of them had apparently died in the explosion or the fire that instantly followed.'

. . . . .

It is a natural protective property of the mind that we cannot visualise such tragedies in a realistic way. We see the 361 blackened bodies only as so many dark blobs. To see what was really involved, we have to turn to a positive case, as for instance in the expressions on the faces of the two miners, shown in Figure 4.1, who were rescued in 1972, against the probabilities, from a fire in the Sunshine Mine at Kellogg, Idaho. There were 91 fatalities at the Sunshine Mine, not dark blobs but real people like those of Figure 4.1. The Sunshine Mine produces silver not coal. The most important use of silver is in the preparation of photographic film, used by environmentalists to photograph birds, and in the making of TV films advocating a world free from pollution.

But to turn now to *Harper's Weekly*:

'It is one of the tenets of the geologist that the lower we go for coal the more gas we strike. The use of electricity in mine haulage has been growing more and more general. Sparks from these electrical

Figure 4.1   Miners rescued from the Sunshine Mine, Kellogg, Idaho in 1972.

appliances quite frequently touch off the gas that is liberated in deep shaft mines.

'In a great many mines the roadways and shafts are exceedingly dry. In these the roadways often consist of coal which has not been taken out. Constant passing to and fro of the system of haulage; the constant passing to and fro of mine employees; the action of the air currents upon the loose particles of coal carried in the cars—all work to charge the air with infinitesimally small particles of coal which are then carried in suspension in the air. So long as the temperature of the air is normal there is absolutely no danger, but let something happen which will raise the temperature in a room, or ignite some of these fine particles of coal dust, and the explosive power of coal is demonstrated in an instant. The explosion seems to travel almost like a prolonged rumble of thunder. It may start from a given point and travel from chamber to chamber until practically the whole mine has been involved in a terrible catastrophe. The result is that mine workers are often burned by the explosion or are killed by the falling material dislodged by the shock. Instantaneous combustion of this kind uses up all the oxygen in a mine chamber, and the men are suffocated because the force of the shock has wrecked both the fan and the power-house and, consequently, has shut off the supply of fresh air.

After one of these explosions comes the noxious product of the discharge, and then the death-dealing work of the explosion is completed. This, in fact, is what occurred at the Monongah Mine. Practically every known device has been used to keep down the dust, to keep the temperature lowered, and to avoid anything that would start an explosion of this kind; and yet, explosions do occur, and are likely to occur, with even greater frequency in the future.'

.   .   .   .   .

We stop the quotation at a mistake. Because of the modern trend to strip mining, the situation has not become worse. Operations open to the atmosphere are largely free from the concentrations of explosive coal dust described in the *Harper's* article. Environmentalists object strenuously to open-cast mining, mostly giving themselves a moral pat on the back when they do so. In effect, environmentalists, while enjoying the benefits of energy from coal, are seeking to restrict the mining of coal to lethal conditions underground, which is not a moral position at all.

Strip mining of all kinds need have little or no adverse impact on the environment. The issue is whether or not money is spent to make a clean job of it, afterwards properly replacing the top soil and replanting in a sensible way. In Britain, where the government has a monopoly of coal production, the cost to British consumers contains a contribution for restoring the environment. In the United States, where coal is mined by private companies, the situation is less simple. Tough U.S. anti-trust laws make it impossible for the different companies to agree a common environmental surcharge among themselves, since any such consultation would be interpreted as a cartel action. This creates a situation in which an irresponsible company that fouls the environment, an irresponsible company whether in the U.S. or abroad, can undersell a responsible company that restores the environment. Sheer commercial survival then forces a situation in which every company is forced into polluting the environment. Slapping environmental restrictions on all U.S. companies, as the Washington regulatory agencies are fond of doing, is no solution to this dilemma, because the restrictions do not apply to companies outside the U.S. Their effect, especially for the primary metals industry, has been to push U.S. companies towards bankruptcy. It is this contradictory situation about which the environmentalists should be protesting.

5
Radioactivity

When anti-nuclear critics are forced into an admission that reactors do not explode, a second string argument often expressed is that, while they may not explode, reactors emit radioactivity. To give substance to this argument, radioactivity is claimed to be the worst of all evils, worse than Flixborough, or than the October air disaster at San Diego, and worse even than the liquid napalm which drenched the campers at San Carlos de la Rápita.

It is true that nuclear reactors emit a small amount of radioactivity (less rather surprisingly than a coal-fired power plant). But everything around us emits a small amount of radioactivity, a rock, a house, a cow. The real issue is *how much* is any particular object radioactive? This is the general question that we shall consider in the present chapter.

From 1960, all British nuclear installations have been required to meet radioactive standards for each individual active substance, with particularly stringent requirements for those like radioactive-strontium and radioactive-iodine that tend to be absorbed into the bodies of animals. The standards were decided by the government on the recommendations of the Medical Research Council.

To make certain that government requirements are met, radioactive emissions from British nuclear establishments are carefully monitored. At Windscale in Cumbria the fall-out is determined in several ways, from the analysis of the milk of cows on neighbouring farms, for example, which immediately shows up the important radioactivities falling from the air on to the grass. The tests are exceedingly sensitive. If anyone anywhere in the world were to cause a nuclear explosion of an appreciable size in the air, it would not need American radioactive sniffer planes to detect the resulting radioactivity. The Windscale cows would do it. The radioactivity found in these tests is typically not more than 10 per cent of the government standard, and indeed for the important substances it is usually about 1 per cent of the permitted levels.

As well as fall-out on the land, it is necessary to monitor liquid effluents into the sea, and in recent years the situation for radioactivity discharged into the sea off Windscale has not been as satisfactory as it has for the fall-out onto the land. For one particular substance the emission has indeed come up to about 40 per cent of the permitted level. The amount of emission into the sea has tended to increase over the years, because of a need to cope with increasing amounts of spent nuclear fuel using facilities that were out of date. Modifications are in progress that will reduce radioactive emissions into the sea to 10 per cent of the permitted level.

The history of the attempt by British Nuclear Fuels to acquire a new processing plant, which is a plant that separates the various radioactive substances in a controlled way from out of a complex mixture, was no recommendation for the then ruling government. The Cumbria County Council, after being 'minded' to give permission, came under pressure to refer the matter to the Department of the Environment. Because of protests, mostly from people resident outside Cumbria, the government then lost its nerve. Unable to make a decision either way itself, an enquiry under Mr. Justice Parker was set up. Justice Parker and his two assessors were required to listen to an enormous volume of evidence, much of it incoherent and repetitive, and some of it imported from abroad ('What's it to thee?' as a Yorkshireman might say.) The facts presented by British Nuclear Fuels had already been made available to Mr. Peter Shore as Minister for the Environment, and to Mr. Tony Benn in charge of the Department of Energy. There could be nothing of a technical nature to be uncovered by the enquiry.

The government was elected to govern, not to pass the buck to Mr. Justice Parker. Mr. Shore and Mr. Benn, with the Cabinet behind them if necessary, should have made the decision. It is true that the actual formal decision on the Windscale reprocessing facility did not lie with the Parker enquiry. The decision lay with Parliament. But once again the government ducked for cover, by making the issue a free vote in Parliament. In the end Mr. Justice Parker and his assessors came well out of it, and so on the whole did Parliament. But not the government.

With this general picture of the emission of radioactive materials from an important nuclear establishment in mind, we come now to the critical question: how much radioactivity is the nuclear industry producing? How much compared with other common sources of radioactivity? Figure 5.1, taken from a recent report prepared by the American Physical Society (APS), gives a summary of the answers to these questions.* The lengths of the various bars measure the bodily damage that we suffer on the average from the various sources. The longest bar, representing the natural background, is made up of three parts, one from fast-moving cosmic particles entering the Earth's atmosphere from outside. The segment marked 'external terrestrial' comes from rocks and soil, and from the houses and buildings that we live in. The third segment comes from the food we eat and the air we breathe, which even under the most natural conditions contain radioactivity.

---

* *Reviews of Modern Physics*, Volume 50, Number I Part II, January 1978.

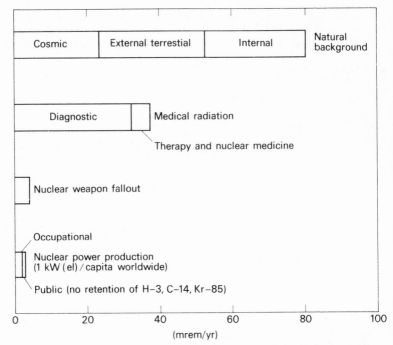

Figure 5.1   Contributions to U.S. average annual whole body dose rate.

The tiny sliver in the lowest bar of Figure 5.1 shows what the public exposure to radioactivity would be if the nuclear industry produced all of the electricity used by the whole world. It is this tiny sliver which is being used as an excuse for trying to stop the construction of all further reactors. Yet if the nuclear industry were to expand sufficiently to meet the need for heat as well, and for the making of liquid fuels, the sliver of Figure 5.1 would not increase even to 10 per cent of the natural background. Public exposure and occupational exposure would still be considerably less than the natural background, and indeed considerably less than medical irradiation.

Figure 5.1 is an averaged situation, and it does not necessarily give a true picture for exceptional cases. Injured and sick people, for example, often experience a far greater degree of medical irradiation than the average for the community as a whole, sometimes considerably more than a hundred times the average of Figure 5.1. In Chapter 7 we shall consider the question of whether in exceptional circumstances the nuclear industry might expose the public to more than the average shown in Figure 5.1. Here we note that the accident of March 28, 1979 to the reactor at Three Mile Island did not lead to exceptional circumstances in this sense. According to the statement referred to in Chapter 3 from the U.S. National Academy of Engineering:

'The exposure to radiation of those living within 50 miles of the plant has been estimated to average about 1 millirem. This is about the amount normally received from natural sources in 3 days of living, or perhaps ⅓ of that received on a jet flight across the country.'

The millirem is the unit shown on the top scale of Figure 5.1. Thus the average dose through the Harrisburg area from the accident at Three Mile Island was comparable to the thin sliver in Figure 5.1. The reader will note the reference in the Academy statement to the three times larger dose received in a jet flight across the U.S., and may well wonder how the latter has managed to escape hysterical comment in the media. The answer is that, so far as we are aware, airline passengers never worry in the least about excess exposure to radiation, and any attempt by the media to raise a scare on this count would simply be laughed out of court. The situation is that scarcely anybody is worried about exposure to radiation on the scale of Figure 5.1 unless it comes from the nuclear industry, and then almost everybody is worried, no matter how small the exposure may be.

6

How damaging is radioactivity?

It is obvious from Figure 5.1 that the radioactivity coming from the nuclear industry is much smaller than the radioactivity we experience naturally. The natural background in Figure 5.1 dominates the average exposure. Now this background is not everywhere the same. In some places the background is considerably higher than the average, in regions where the rocks and soil contain exceptionally large concentrations either of uranium or of thorium—the other naturally radioactive element. An example of a place where the general run of uranium in the rocks is higher than normal is the county of Cornwall, and in the United States there are many regions of Utah, Colorado, and Wyoming where this is so. Yet we see no marching in the streets, no demands that everybody be instantly evacuated from these exceptional places. In India there are regions where the background goes to about ten times the average of Figure 5.1. Instead of the background being represented there by a bar about three inches long, the bar would be about a yard long. Yet stacks of dead bodies are not to be found in those parts of India, no stacks like the ones at San Diego and San Carlos de la Rápita. To produce a situation of comparable gravity to these latter disasters, it would need radioactive doses represented on the scale of Figure 5.1 by a bar *several hundred yards long*.

In order to understand these issues more clearly, let us consider the situation from a slightly more technical point of view. Radioactive materials emit fast-moving particles, which in the early years of the present century were known as either α or β particles. Later work showed that α particles were really the nuclei of helium atoms, and β particles were really electrons. The so-called γ particles, also emitted by some radioactive materials, turned out to be radiation like light, only of very short wavelengths. The X-rays used by the medical profession are also radiation, with wavelengths intermediate between γ-rays and ordinary light. All these forms produce a trail of dead or damaged cells if they enter the body of an animal. Collectively they are called 'radiation', and an animal exposed to any of them is said to be irradiated.

The trail of damaged cells can be thought of as an exceedingly minute pinprick, but a pinprick that may go deeper into the body than an ordinary pinprick. It is because the pinpricks caused by radiation (α, β, γ, or X) are so tiny that we do not feel them, although they are happening all the time from natural causes.

A pinprick is a mild kind of wound, and a single tiny pinprick from radiation is an exceedingly mild wound. The problem with a high dose of radiation is that it causes a multitude of such mild wounds, which collectively may add up to something serious. Just as the body immediately moves to repair the damage done by an ordinary wound, so it moves to repair the many fine pinpricks caused by radiation. Since time immemorial the bodies of animals have been repairing the radiation caused by the (not-so-pure) environment.

We can now understand a little better the meaning of the scale used in Figure 5.1. It is a scale of damage, a measure of the number of pinpricks which the various sources of radiation cause. What has been done in drawing up the figure is to work out the damage caused by each form of radiation, for the top bar of the figure the forms of radiation in the natural environment, in the second bar the radiation administered by the medical profession, in the third bar the radiation from the military nuclear explosions which occurred in the air before such explosions were banned by international treaty. The very small bar at the bottom of the figure shows the damage coming from the civil nuclear industry.

If we do not feel the tiny pinpricks caused by radiation of all forms, why should we worry about them? Because they might cause other things that we do feel. The crucial question is therefore: How much in the way of pinpricks causes what in the way of other more serious things? There *should* be a decisive answer to this question, but unfortunately there isn't. The blame for this lack of knowledge lies, not with the nuclear industry, but with the medical profession. Only the medical profession is permitted to expose people deliberately to radiation. For fifty years or more the public has been exposed to X-rays for diagnostic purposes. Sometimes the doses have been very large, hundreds of times the bars of Figure 5.1. If X-ray doses had been quantified, and if the subsequent medical histories of patients had been investigated, the information necessary to answer our question would long ago have become available.

The only worthwhile information presently available is at very high dose levels, of the order of a *thousand times greater* than the bars of Figure 5.1. The information comes from a careful follow-up of the medical histories of survivors from the explosion of nuclear bombs over the Japanese cities of Hiroshima and Nagasaki. Next to the 'conventional' air attack on Dresden, which killed 135 000 between February 13 and 15, 1945, the nuclear attack on Hiroshima, which killed 90 000 on August 6, 1945, is the best-remembered bombing raid of the Second World War. Older people who recall photographs of the damaged condition of survivors from Hiroshima and Nagasaki will be surprised to learn that the life expectation of survivors has been slightly but significantly better than that of a comparable control group among the normal Japanese population.

We mentioned in Chapter 2 that British traffic kills about 7000 each year, and that the annual total of persons injured is about 350 000. Unlike the Japanese survivors, it is well-known that among those injured by road traffic are many unfortunate cases who are by no means able to resume a normal life, and for whom the life expectation has been seriously shortened.

When the outcome of the investigation of Japanese survivors was published (S. Jablon and H. Kato, *The Lancet*, November 14, 1970) the result was challenged, apparently in the hope that the condition of survivors would on reinvestigation turn out to be less favourable. It has not done so, and the efforts of the anti-nuclear movement are now directed towards explaining away what for them is a highly inconvenient result. Unlike survivors from wounds of a normal type, it seems that survivors from large doses of radiation, represented by bars a hundred yards or more long on the scale of Figure 5.1, do not have their future expectation of life much impaired, if indeed it is impaired at all. Nor have any genetic effects been found in the progeny of the Japanese survivors.

This is not quite the end of the matter, however. The Japanese experience showed that there is one particular disease, leukemia, for which the chances are increased in the years following exposure to very large radiation doses. Thus at a dose 1000 times the uppermost bar of Figure 5.1, i.e. for a bar about 100 yards long on the scale of this figure, the chance of subsequent death from leukemia was found to be about $\frac{1}{100}$ per year, comparable to the lung cancer risk incurred by heavy cigarette smokers.

How are we to translate this information for very big doses into the much smaller doses shown in Figure 5.1 that we all receive in our daily lives? If the many small wounds coming from radiation exposure act independently of each other in their effect on the chance of contracting leukemia, then we could say that at $\frac{1}{1000}$ of the dose the chance would be one-thousandth of $\frac{1}{100}$, i.e. $\frac{1}{100000}$ per year for the situation of Figure 5.1, with the nuclear industry contributing a chance of less than $\frac{1}{1000000}$. But if the many small wounds coming from high radiation exposure do not act independently of each other, if it is the concentration of many wounds together that has a significant effect, as many medical authorities think may be the case, then the effects of the low doses of Figure 5.1 would be negligible so far as leukemia was concerned.

The missing component in our knowledge comes at just this point. We do not know for sure which of these two possibilities is correct. What the nuclear industry therefore does, both in respect of its own workers and the public, is to assume the worst case, the first of the above two possibilities, which is often referred to as the 'linear hypothesis'.

Inevitably, there are emotionally committed persons who have arranged such fragmentary data as exists in such a way as to claim that the linear hypothesis is *not* a worst case assessment. Even in simple everyday situations, it is often possible to arrange data so as to arrive at incorrect conclusions. Suppose one examines the registration numbers of a hundred cars in a typical parking lot. Since registration numbers are usually issued in sequence, all cars—millions of them—would show the ten separate digits 0, 1, 2, 3, 4, 5, 6, 7, 8, 9 with essentially equal frequency. But an observer examining only a hundred cars would hardly ever find the digits equally distributed. Through natural statistical fluctuations some would appear more often than the average and some less often, with deviations that at first sight seemed quite substantial. If each registration number had four digits, for a total of 400 digits over the hundred cars, the average for each digit would be 40. But it would not be at all unlikely for one particular digit to show up at what is called a two-standard-deviation level, which would be 52, twelve above the average. If the exceptional digit happened to be 7, the observer might then make the false claim that the parking lot in question had the astonishing property of attracting cars which happened to possess the registration digit 7.

If our observer were to visit several such parking lots he would of course find similar effects everywhere, but with other digits being enhanced elsewhere. Some lots would apparently favour 0, some 1, some 2, and so on. This finding should convince the observer that the claimed effect was spurious, but if the observer were emotionally committed to the belief that particular parking lots attracted particular digits he might even seek to emphasise his claim, since after visiting a score or so of such lots he would be likely to find a still larger fluctuation of some digit, extending up to three standard deviations (a count of 58 in the above example) which the unwary can often be browbeaten into accepting as a valid level of statistical significance.

Exactly the same procedure can be, and has been, adopted to analyse the occurrence of cancers in small sample populations that have been exposed to low radiation doses. The population could be chosen for excesses in the natural background of Figure 5.1, as for people living in Aberdeen, or on Dartmoor, or on certain of the Cornish granites. They could be people who have been exposed to medical irradiation, or people who have worked in the nuclear industry. Instead of numerical digits, one looks now for the various forms of cancer, which are numerous enough for one or other of them to turn up at about the two-standard-deviation level, simply as a matter of chance. And if ten to twenty such population samples are examined it becomes likely, again as a matter of chance, that in one or other of the samples some form of cancer will be found at a three-standard-deviation level. This particular special case is then emphasised as being of 'serious' statistical significance. The other population groups with smaller deviations are forgotten, and discussion is muted of all the other forms of cancer that were looked for. In such a way false evidence of deleterious effects at low radiation doses can be produced. Similar false evidence could be produced for population samples chosen by any criterion one pleased.

We turn now to what may be a genuinely disturbing situation, but a situation that has nothing at all to do with the nuclear industry. The uppermost bar of Figure 5.1 gives the average bodily damage that we all experience from the natural background. Contained in the natural background is radioactivity from the heavy gas radon, not produced in fission, but produced by the natural decay of uranium in the ground. Radon is especially dangerous because it emits the worst kind of radioactivity, α particles, and because, since we breathe it in the air, it acts specifically on the lungs. The APS report mentioned in Chapter 5 finds that the resulting situation for the lungs is up to 5 times worse than the average for the whole body. Over a whole lifetime, the lungs experience a damage not much less than the damage experienced by many of the survivors of Hiroshima and Nagasaki.

It has for long been difficult to believe that cigarettes can be the sole cause of the high risk of lung cancer to which heavy smokers of cigarettes are exposed. An additional effect seems to be needed. The additional effect may well be radon, with the smoking itself having the subsidiary influence of interfering with the body's natural damage-repairing process. If smoking has the effect of *accumulating* the radon damage, of preventing or distorting the repair mechanism, then the high risk for smokers becomes understandable.

So too does the exceedingly high mortality of coal miners from respiratory diseases, in Britain 35 times the norm for the general population. Breathing coal dust, and cigarette-smoking as many miners do, must of course have a deleterious effect on the lungs, but to the extent of a factor 35 seems doubtful. Radon, aggravated by coal dust and smoking, is a far more plausible possibility. Mine shafts are a principal source of radon, with the gas thus afforded exit points from the underground regions where it is generated. For miners, spending some 40 hours a week underground, the dangers are only too obvious.

Here again is an important, and perhaps crucial, reason why mining should if possible be always open-cast. So long as society is vitally dependent on mining, no environmentalist worth a grain of moral salt should object to strip mining. Instead of protesting about such activities, the genuine environmentalist should be prepared to pay gladly for restoring top-soil and replanting at the end of such operations; pay his or her share, that is to say. The cost should not be paid for by the lungs of miners.

Another point concerns the positioning of mining towns and villages, often close to a multitude of mine shafts. 'All the local people knew about the hollow hills. The Consolidated Coal Company of Boston had honeycombed them with mine shafts until it seemed the town might one day fall in.' This is a perfect description of conditions ideally suited to the emergence of radon from the Earth, and a description that could be applied in thousands of places besides Monongah, W. Virginia. It is about such situations that the marchers of Figure 3.1 should have been protesting. The gas masks they were wearing would then have been appropriate.

7

Nuclear accidents

Energy exhaustion would cause the early deaths of 4000 million people, as we pointed out in the beginning. Compared to such a fantastic disaster, the deleterious effects of the radioactive emissions from civil nuclear reactors, shown in the lowest bar of Figure 5.1, are minuscule.

Faced with this situation, the anti-nuclear movement would have little chance of carrying the public if the truth were admitted. They have instead invented a kind of imaginary radioactivity, which has been used to delay the construction of reactors, with the inevitable consequence of driving up capital costs, and so of increasing the eventual cost of electricity to the public.

In the face of the overwhelming evidence that reactors cannot explode, the concept of an-explosion-but-not-an-explosion has been introduced. While it is admitted, even by such anti-nuclear publications as *The Guardian*, that reactors cannot explode in the sense of a nuclear bomb, it is claimed that reactors *can* explode in the sense of spewing out radioactivity over the surrounding countryside in such quantities as to increase greatly the lowest bar of Figure 5.1. In this way 'imaginary radioactivity', radioactivity that is *not* being spewed out, is invented.

There has been just one case of a nuclear accident in Britain in which radioactivity was emitted in an appreciable amount into the public environment. It occurred as long ago as 1957, and it happened to a military not a civil reactor. The circumstances, and the consequences that ensued from it, are described by Professor S. E. Hunt in his book *Fission, Fusion and the Energy Crisis.* *

'The release occurred when the reactor was in the shutdown state and was due to a phenomenon in which ... carbon atoms in the graphite moderator (are) displaced from their normal position of lowest energy. The graphite consequently stores energy when the reactor is in operation and releases it in subsequent heat cycling, as the carbon atoms return to their normal positions. In this particular case the stored energy was sufficient to burn some of the graphite and to melt part of the fuel, and since this reactor was air cooled and not equipped

* Pergamon Press, 1974, p. 58.

with the now standard containment, the fuel melting led to the escape of a certain amount of radioactivity, mainly radioactive-iodine, to the surrounding countryside. The level of contamination did not represent a hazard to life, but was such as to make the milk production in an area 10 miles by 30 miles unusable ... due to the accumulation of radioactive-iodine by grazing cattle. Fortunately this fission product nucleus has a half-life of only 8 days and the situation returned to normal within a few weeks.

'Although the Windscale incident was the only case of an escape of radioactive material from a reactor in this country (Britain), it led to an increase in the severity of the ... vetting of reactor designs, siting and operating procedures. In examining reactor proposals, great imagination and ingenuity are involved in postulating the so-called "maximum credible accident", where it is assumed that a combination of several unlikely and potentially hazardous faults occur simultaneously, and the reactor design and operational procedure must be adequate to prevent an escape of radioactive material under this combination of adverse circumstances. In the author's experience the crashing of an aircraft on a reactor building, simultaneously with two other only slightly less unlikely fault conditions, were combined to constitute a "maximum credible accident" against which a reactor had to be safe.'

We would add that this incident in 1957 also played another significant role, by inducing the government to set up the radioactive standards that were discussed in Chapter 5. The concept that a reactor must be designed to prevent the escape of radioactivity, even though an aircraft crashes into it at the very moment when two other only slightly less unlikely independent internal faults occur, was introduced by the nuclear industry long before anti-nuclear protest movements gathered strength. These protest movements have appeared only *after* the 1973 oil crisis made the need for nuclear energy both necessary and urgent for all the developed nations*. The careful self-policing of the nuclear industry, described by Professor Hunt, occurred long before 1973.

There is almost no limit to the safety precautions that one can take in regard to the simplest operation. One could prepare for an hour thinking of safety measures before walking down a flight of stairs. What the anti-nuclear movement has done is to insist to politicians—concerned with little else than vote-gathering—that every precaution anybody can think of must be introduced into the design of reactors. It is commonplace nowadays for features to be incorporated into reactor designs in order to cope with a situation that is not likely to occur once in a million years. The National Radiological Protection Board (NRPB) published in September 1977 a study in which the chance of a postulated large

* It is an interesting exercise, for example, to compare issues of the *New Scientist* for 1972 with those for 1977. Rather little that was anti-nuclear in 1972, a strident crescendo of articles and comments in 1977.

emission of radioactivity into the environment was said to be much less likely than one in a million years of reactor operation, 'and perhaps as small as one in a thousand million years of operation'. The effect has been to load designs with an accretion of unnecessary features, doubling construction times from what should sensibly be about five years to ten years or more. Capital costs have therefore spiralled, and since capital costs represent the biggest item for nuclear electricity, it is the public that will eventually pay for the activities of the anti-nuclear movement, as well as suffering from the energy shortages that it will promote.

While the NRPB's study was worth making as an abstract exercise, the attempt to attach a probability to it was absurd. For events that have never happened in human experience, it is impossible to assign probabilities in a meaningful way. So far as anything known to the study group was concerned, a chance 'perhaps as small as one in a thousand million years' was indistinguishable from a chance of zero. In any case it is likely that within a few tens of thousands of years the U.K. will find itself once more in a full-blown ice-age, with Windscale and Westminster buried under a glacier hundreds of feet thick.

The anti-nuclear movement is now campaigning for the publication of what is called an incident count. This is to be a device for pretending that an accident has occurred when an accident has not occurred. An 'incident' is an occasion on which a safety feature has been brought into operation, as for instance a motorist might avoid trouble on the road by pressing the footbrake of a car. A quick calculation shows that in the course of a year there are several thousand million such 'incidents' on British roads. Are we then to argue that the annual total of road accidents runs to thousands of millions?

The correct way to assess the probability of future accidents is on the experience of past accidents for similar circumstances. For the civil nuclear industry there has been no clear-cut evidence of any accident leading to death or injury, which means that according to every extrapolation formula known to mathematicians the best estimate for the future number of injuries and fatalities is zero. This is not to say that zero will necessarily turn out to be correct, but zero is the best technical prediction that can now be made. Should injuries and fatalities really occur in the future, experience will then permit improved estimates to be made. Attempting to guess what has never been shown to happen is not merely logically unsound. It is a sure recipe for the dissemination of alarmist propaganda of every kind.

There was no death or injury list from the accident at Three Mile Island near Harrisburg, Pennsylvania on March 28, 1979. Nor did it cause a fire, as the accident 22 years earlier at Windscale had done. Nor was a serious quantity of radioactive-iodine emitted into the environment. The *Daily Telegraph* for April 9, 1979 reported:

> 'Health officials in Pennsylvania and neighbouring New York State said yesterday that some radioactivity had shown up in milk since the radiation leaks from the crippled nuclear power plant.

'The prevailing winds carried some traces of radioactivity over the farming areas, but officials in both states insisted that they were not "overly concerned" because tiny amounts of radioactive-iodine found in milk were only slightly above the threshold of detectability.'

Instruments for detecting radioactivity are exceedingly sensitive. They can detect radioactivity far below the level at which there is the slightest danger. For the amount carried by the 'prevailing winds' to have been only slightly above detectability, the leak of radioactive-iodine at Three Mile Island must have been minuscule indeed.

The accidents at Windscale and at Three Mile Island differed greatly in their causes. The overheating of graphite at Windscale was a quite subtle effect, new in 1957 to most physicists and nuclear engineers. The situation at Three Mile Island was anything but subtle. The situation there was described by David Fishlock in the leading article of the *Financial Times* for April 30, 1979:

'... it can be inferred that the operators probably made six serious errors—three of them within 15 minutes of the first sign of trouble. But the first, for which they may be in breach of their conditions of licence and thus open to legal action, was to have a crucial emergency feedwater valve closed which the safety officials stipulate must always remain open while the reactor is operating. *In fact, it had been closed for two weeks before the accident.*

Then, in quick succession, they failed to recognise that the safety valve whose whistle first aroused local residents had failed to close properly, even though this was a fairly common occurrence with Babcock and Wilcox reactors. They failed to notice that water they were pumping into the reactor was overflowing into the containment, carrying with it radioactivity. And they turned off a pump whose job is to provide emergency cooling to the reactor.

In the next hour or two the operators failed to close a blocking valve, so the system continued to lose its cooling water. Steam built up in the cooling circuit so that, as one observer put it, "the pipes in the place were all vibrating like mad". Then the operators made what was probably their most serious blunder. They turned off ... (all) the pumps, depriving the reactor of its main feed of coolant.'

The term 'human error', which has so far been used to describe these events, scarcely seems appropriate. Rather can one say that, if the aim of those concerned with the reactor had been to discover its behaviour under conditions of extreme misuse, they could hardly have devised a more effective scenario. What happened inevitably was that the fuel elements became hot, but there is no evidence that even one fuel assembly experienced meltdown, and the reactor core is now believed to have been less severely damaged than had been forecast. Concerning the fuel and the core David Fishlock writes:

'Nuclear fuel in this type of reactor consists of pellets of a ceramic—uranium oxide—with very high melting point, packed into long vertical tubes of zirconium alloy. These tubes, hanging vertically in the reactor core, are the first of *three* lines of defence* of the reactor against the escape of radioactivity into the atmosphere. The low coolant water level within the reactor allowed the upper ends of the zirconium tubes to come into contact with the hot, high-pressure steam circulating in the system. They reacted chemically, dissolved and allowed fuel pellets to be scattered around the reactor. This produced copious volumes of hydrogen gas, to form a big bubble which—unlike steam—could not be condensed back to liquid and pumped out of the reactor.

'The problem here was never, *as was widely reported*, the danger of an explosion that might blow up the reactor vessel and breach the *second* line of defence against a (serious) escape of radioactivity. Such an explosion within the pressure vessel, the engineers concluded, could not possibly occur. Their problem was how to bleed off a lot of hydrogen quickly from a system for which the contingency had never been expected, so that cooling could be fully restored.'

Engineers and scientists conversant with the nuclear industry are agreed that the most serious aspect of the technical situation was the failure to expect the formation of the hydrogen bubble. The better aspect of the technical situation was that, in spite of the gross abuse to which the reactor had been subjected, its second and third lines of defence were never required. Indeed, in retrospect it is clear that the reactor itself came quite well out of problems that had arisen so unnecessarily. The disturbing element still lies in the circumstances that led to the emergency. Certain of the 'blunders' described by David Fishlock can be understood in terms of panic decisions taken in a difficult situation, but others cannot be so understood. Since the whole world has been affected by Three Mile Island, much concerning the causes of the accident remains to be explained by Metropolitan Edison, and by the Government of the United States. The special report commissioned by the President himself is being keenly awaited. Especially would one like to know exactly how it came about that the crucial emergency feedwater value was closed two weeks before the accident, and why a warning light (such as we all have in our cars) was not seen by the operators.

---

* The second line of defence is the containment wall of the reactor itself, and the third is the thick concrete wall of the reactor building, the wall we see from outside.

8

The day the dam broke

A person who measured, not too accurately, the lengths of the days around mid-July, would find them to be hardly changing. Such a person might then argue that winter would never come. By mid-September, however, the days shorten perceptibly, and the coming of winter becomes only too clear.

So it is with reserves of oil. For Americans, stampeding nowadays to import more and more oil, the time is mid-July. But by 1990 the world will be in the September of its oil reserves. By then, with the prospect of an energy crisis clear to everyone, it is likely that the anti-nuclear movement will have gone to ground, and will be as hard to dig out as moles in a hard-frozen winter. But the damage they have wrought will live on after them. The nuclear reactors that will be needed desperately in the decade 1990–2000 should either be under construction, or in the planning stages, now. Some are, but many are not. For a variety of reasons, of which the marching and shouting in the streets is the least, reactor construction is proceeding only very slowly. Delays caused in the United States by legal harassment, and throughout the western democracies by the intimidation of politicians at the polls, are leading to the spiralling costs which are causing electricity companies and authorities to shy away from the nuclear programs they would otherwise like to develop.

The inevitable outcome of these delays will be energy shortages in the future, shortages that will somehow have to be bridged for a while by energy sources other than nuclear. In this chapter, and in the next three chapters, we shall take a look at some other sources of energy, beginning here with hydroelectric power.

For countries with high mountains and small populations, as for example Switzerland, Norway, and New Zealand, hydro-power can validly be a principal source of energy. But for the world as a whole hydro-power, however enthusiastically prosecuted, cannot provide more than a few per cent of the needed energy.

People generally are surprisingly relaxed about the dangers of hydro-power, and as we saw in Chapter 2 have been willing to accept a considerable measure of visual pollution of the environment in its development. If the same standard of risk-accounting were adopted for hydro-power as is used for nuclear power, hydro-power would immediately be ruled out of court, for two reasons. It would be seen as a fatal flaw that hydro-power involves the damming of a huge volume and weight of water high above the places where people live. The second flaw is more subtle. Immediately after its construction, a dam is usually in a safe and sound condition. As time goes along, however, a dam is called on to take repeated shocks caused by natural events, which could be unexpectedly large floods, earth-movements or earthquakes, or flashes of lightning repeatedly striking the main retaining wall. The trouble is that as a dam weakens there is not much that can be done about it, short of taking down the existing dam and building a new one. In most of the many examples of dam disasters subsequent enquiries have usually revealed that the developing weaknesses were known, and that the big job of constructing a new dam had been put off until it was too late.

It may have been the Johnstown Flood of May 31, 1889 that gave James Thurber the idea for his famous drawing of 'The Day the Dam Broke'. To quote from Welman and Jackson's *Disaster Illustrated* (*ibid* p. 14):

'He came riding down the mountain like something out of a nightmare. "The dam is going. The South Fork Dam is going!"'

'The panic in civil engineer John G. Parke's voice alarmed the folks in South Fork, which lay just two miles below the dam. Every spring it seemed someone came galloping down the mountain, yelling that the Conemaugh Lake Reservoir was about to dump its water into the valley below.

'It never happened. The first few times, everyone as far down as Johnstown, 16 miles away, had run for the nearby hills. But not anymore.

'This time, however, there was something in John Parke's voice that sent the population of South Fork heading for the hills.

'John Parke reached the village of South Fork just before noon, and immediately sent two men to the telegraph office to warn the towns and cities down the valley. Years later, some would claim that Conemaugh and Johnstown and the other towns and villages never got any warning because the telegraph wires were down as a result of torrential rains that had plagued the area for a full week. But as *Harper's Weekly* put it: "Thousands of people discredited the alarm because it was like the false warnings they had heard before."

'Up in the mountains, the South Fork villagers had begun to wonder if it wasn't another false alarm. The dam was leaking, but nothing else seemed to be happening. Suddenly, a section of the huge dam seemed to move off, and a 75-foot wall of water roared into the valley. South Fork and the little towns beyond disappeared in the twinkling of an eye, and the great wall of water surged towards Johnstown at 40 miles an hour. Squeezing through a very narrow pass at Woodvale, just above Johnstown, it reared up to 90 feet and with a mighty roar fell like a judgment on the town.

'Those who could do so ran for high embankments or the hills. Many were caught unawares in their homes, some were stunned and unable to move, and still others were too burdened with children and loved ones to move fast enough.

'From precarious footholds on high ground, people watched in horror as the whirling waters destroyed their town and tossed their neighbours about in the debris. They tried to help, holding out straps, ropes, or hands to drag anyone they could to safety. But the job soon became hopeless. From South Fork to Johnstown, more than 2,200 died.'

Looking at the date of the Johnstown disaster, 90 years ago, one is inclined to think it couldn't happen nowadays. Perhaps not with quite the same details. But almost every year a dam collapses somewhere or other. The reader may recall a serious disaster in N. Italy, and to quote Welman and Jackson again (*ibid* p. 14), this time for the year 1972:

'In the Black Hills of South Dakota the weather report was dreary, but not alarming. Variable cloudiness was predicted for June 9, with chances of scattered showers and thunderstorms. But on the ninth, the scattered showers became a steady downpour the likes of which Rapid City had never seen before.

'So unusual was the prolonged deluge that the residents of Rapid City didn't know enough to be worried, and were totally unprepared for what was to come. In 6 hours, 14 inches of rain fell. The National Weather Bureau reported that the atmospheric conditions over the area on June 9 might occur only once in a hundred years.

'By 7.15 p.m., the local weather station was issuing high-water warnings. Rapid Creek, the pretty stream which wound down from the hills and which gave the city its name, was disturbingly full. However, it wasn't only the creek that troubled Mayor Dan Barnett. His worry was Canyon Lake, which lay above the city's most desirable residential section. A large, man-made lake that straddled the creek, it was held in place by a dam that was due for its five-year overhaul later that summer. Since it had been a wet month, Barnett wasn't sure the dam would hold against this extraordinary rainfall. He got together with the chief of police, the director of public works, and a crew to inspect the condition of the dam.

'At 10.00 p.m., the group received a call from a man who lived by the creek, up above the lake. He'd just seen a "wall of water" coming down the creek, and it would hit the lake in about 20 minutes.

'The Mayor called the local radio station to issue announcements to abandon the area at once. He rallied the National Guard to evacuate people. He ran up and down the creek himself, asking people to leave. They all looked at him as if he were crazy. "They wouldn't listen," he said in amazement afterwards.

'Between 10.44 p.m. and 11.00 p.m., the dam broke. With a terrible roar, the wild wall of water hit the city and turned its streets into turbulent rivers. Frantic people ran for their cars, only to have them quickly fill with water, cutting off all means of rapid escape. Houses sailed down the street with people clinging to the rooftops. National Guardsmen did their best to help, and one was swept away in the fierce tide as he let go of his truck door in an effort to save a little girl. Others watched helplessly as a house with a roof full of people floated down the creek and struck a bridge, forcing everyone into the churning waters.

'Those lucky enough to be on, or in, standing houses were as helpless as the National Guard to help less fortunate victims. In nearby treetops, they could see children trying to hang on for dear life. Some of them were rescued, but not all. "Our daughter just floated away," one broken-hearted father reported afterwards. "She tried holding on to a tree, but she just floated away."

'All power failed before 2.00 a.m. There was no light, no communication. Nobody knew what was happening outside their own little realm of horror. It wasn't until 6.00 a.m. that any of them knew the worst was over, when someone managed to broadcast on a citizen's band. The first message received was not cheering: "Stay in your homes. Do not impede emergency-vehicle traffic. Do not drink the water. Boats are needed immediately. If you find a body, do not move or touch it."

'Before the day was out, it was obvious that this was South Dakota's biggest disaster. Senator George McGovern interrupted his campaign for the Democratic nomination for President and flew to the stricken city, where he waited with fellow citizens to hear the sad results of the flood.

'There had been a lot of campers in the area, out in the nearby woods of the Black Hills. What had become of them? "They'll be picking them out of there all summer," sighed the chief of police.

'By the end of the summer the official count was 236 dead, with 124 missing. In Rapid City, they know one thing now for sure. When you live on a flood plane—like the one that exists below Canyon Lake—and someone tells you to get out, you get out.'

In retrospect, the behaviour of people in Johnstown and Rapid City may appear to have been foolhardy. But their behaviour was really normal, just as the behaviour of the people of Pompeii in AD 79 was normal. What is emphatically not normal is the behaviour of anti-nuclear activists in raising a great storm over minuscule risks and over things that may never happen 'once in a thousand million years'. Notice from the above account that dams can be destroyed by events that happen once in a hundred years (according to the remark of the Weather Bureau) not once in a million years as with nuclear reactors. If the nuclear standard of a 'maximum credible accident' were introduced into hydro-power, effectively all dams throughout the world would have to be immediately dismantled.

As this book goes to press, the news has come through that the 7-year-old dam in Gujarat, W. India has collapsed and killed some 2000 people, a disaster comparable to Johnstown. It is indicative of the neurotic unhealthy state of the American media that, whereas *Time* devoted a considerable fraction of a whole issue to the minor affair at Three Mile Island, it could manage only 25 lines on the Machhu dam.

9

How stands solar?

A few months ago with thick snow outside the window, with an icy wind blowing from the Arctic, solar stood absolutely nowhere. Yet one can argue that the energy requirement of society is a mere $\frac{1}{100}$ of 1 per cent of the energy of all the sunlight coming through the Earth's atmosphere to ground level. The human need for energy is apparently so small compared with the vast bounty provided by the Sun that the non-technically minded person can be pardoned for believing in the eventual emergence of solar power as the saviour of future generations. Even if an adequate solar technology does not exist today, developments in the coming decades and generations will surely suffice to solve practical problems. So it is argued by many who favour so-called 'soft' energy paths.

It is easy to be confirmed in this view by articles appearing with considerable frequency in reasonably serious scientific literature, such as the following which appeared over recent months in the *New Scientist*:

BRIGHT OUTLOOK FOR THIRD WORLD SOLAR POWER
22/29 Dec., 1977
SUNNY FUTURE FOR POWER SATELLITES   25 May, 1978
SOLAR ENERGY HAS A BRIGHTER FUTURE   29 June, 1978
SOLAR ELECTRICITY SPEEDS DOWN TO EARTH
30 Nov., 1978

The editorial policy of the *New Scientist* is clearly to provide articles on solar energy with glowing enthusiastic headlines, and just as clearly to damn articles on nuclear energy with headlines designed to arouse the reader's apprehension.* The following contrast between solar and nuclear power is the one which the *New Scientist*, as a *science* magazine, should really emphasise to its readers:

1 New-style solar power, excluding wood, provided the United States in 1977 with much less than 1 per cent of its energy requirement.
2 In the severe winter of 1977/78 there was a near-breakdown in the

---

* This statement may be verified in almost every issue of the *New Scientist*. A general review of energy developments in the U.K. by Peter Chapman (15 July, 1976) was headed *The Nuclear Explosion*. The article had nothing to do with nuclear explosions.

U.S. electricity supply, due in a considerable measure to the freezing of coal dumps into immovable mountains of solidified material—the lumps of coal stuck tenaciously together. In this unexpected situation the nuclear industry continued to work normally and in the mid-West accepted additional load.

Not to be outdone in the matter of headlines, *Time* for May 1, 1978 carried an article entitled *The Sun starts to shine on solar* in which the diagram shown here in Figure 9.1 was given. After examining this figure, the reader may wonder what the talk of an 'energy crisis' is about. If a small underground storage tank, together with an apparently simple gadget on the house roof, can solve heating problems in the home, where then is the crisis?

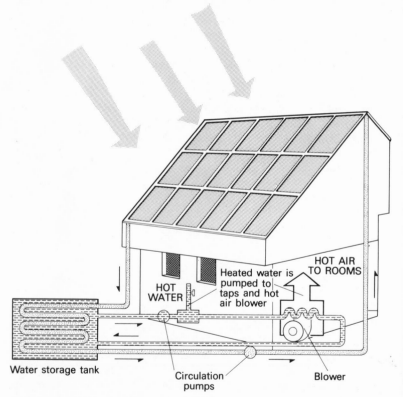

Figure 9.1   A domestic heating system fuelled by solar power.

A friend fitted the device of Figure 9.1 for what seemed to us a sensible purpose, to heat a swimming pool. Since our friend lives in California, where there is almost a surfeit of sunshine, we expected the system to work. It did, but in summer only, and only if you were content to take your swim in the mid-afternoon. Our friend, a distinguished scientist, made his purchase after a careful examination of a number of installations available on the market, so we doubt that his experience was any worse than normal.

For a person anxious to purchase the system of Figure 9.1 we offer a few remarks. The roof of an existing house should obviously be tested carefully to make certain that it can bear the weight of the liquid-filled solar panels. One should also make quite sure that the system will not be ripped away by the strongest wind that is likely to blow in a ten- to twenty-year period. Extreme gale force winds occur several times a year in the north of England, where we live, and every few years there are speeds in exposed locations above 100 m.p.h. A high wind in November 1978 snapped the pine tree shown in Figure 9.2 like a matchstick, although the tree was in good condition with a trunk diameter at the snapping point of more than a foot. The penalty if one lives in a remote spot might be no more than the loss of the installation, and of the house roof. In a well-populated area, there would be the possibility of heavy

Figure 9.2 A mature pine tree snapped by high winds.

bric-à-brac blown from the installation causing serious injury to others, in which case legal suits for damages could be expected to follow.

It is important to realise that solar energy is most available when it is least wanted, in summer; and solar energy is least available when it is wanted most, in winter. The warm water needed on a cold winter day cannot normally have been heated the same day, or even heated the previous week or month. Warm water for winter use must be heated in summer, if an installation is to be satisfactory. This means that the warm-water storage tank must retain heat for about six months.

In our experience even well-lagged domestic hot water tanks within the home, not in frozen ground outside, go cold in four or five hours, once the heat source is cut off. Increasing the size of a tank, and increasing the thickness of the lagging, increases the cooling time, however. With extreme care to prevent heat losses, and with a storage tank the size of a swimming pool (not the small tank shown misleadingly in Figure 9.1), it may be possible to retain warm water for some thousands of hours, as is required. Yet a tithe of the effort needed for such a project, applied to insulation within the home itself, would cut fuel consumption in a marked degree. Even for older houses with awkward design features we ourselves have managed through attention to insulation to cut fuel requirements to less than a half of what they formerly were.

The system of Figure 9.1 delivers energy in its lowest grade form, as warm water, not even as really hot water. Most of the demand by society is for energy in much higher grade forms than this. Of the higher grades, electricity is usually considered the most convenient. At the standard of living enjoyed in the U.S., and towards which all the developed countries are aspiring, the per capita requirement for high-grade energy in all its forms is about 7 kilowatts (kW), so that a family of four requires some 28 kW. Let us see what kind of a solar device might be constructed to meet this higher requirement.

The solar power incident over an area of 1 square metre (1 m²) is 1.4 kW, provided the area is turned directly at the Sun, and provided it rides freely out in space. The power incident over an area of 20 m² out in space would thus suffice for our family of four, if it could all be turned into electricity, and if the electricity could be made available at ground level. Sunlight can indeed be converted into electricity, by a device known as a solar cell, which consists essentially of a sheet of silicon, and the electricity so produced can be converted at high efficiency into all other forms of energy.* The first snag, however, is that solar cells are both inefficient and expensive. The efficiency at present is about 10 per cent, but there are hopes of an eventual increase to 20 per cent. Likewise there are hopes of a sharp reduction in cost, hopes which lead to optimistic headlines for solar power such as were quoted above.

Let us suppose these hopes realised, in which case at 20 per cent efficiency the required collecting area for a family of four would be 100 m², not 20 m². If, instead of a flat sheet of solar-cell material with this

* Even into a liquid fuel.

large area, a focusing mirror were made, the problem of collecting the resulting electricity efficiently would be eased, and this too we will suppose to be done.

While there are projects, still undeveloped, for actually collecting solar energy from space, in both the near term and in the intermediate term it is necessary to think practically of collection at the Earth's surface. Assuming, favourably, the mirror to be steered in daytime so as to follow the Sun, the system would be operative during winter months for about one-third of the time, because winter nights are about twice as long as winter days. To have adequate power for the winter months it would therefore be necessary to increase the estimate of 100 m² for a station receiving continuous sunlight in space to 300 m² for an installation at ground level. Obscuration of the Sun by cloud, and absorption by atmospheric gases and particles, would usually reduce the efficiency still more, to about 50 per cent, so that the necessary area would have to be increased further to 600 m². Nor is this quite the end of the matter. It would be necessary to buffer the electricity supply, as for instance with the help of storage batteries. The efficiency of such a buffer store would be about 75 per cent, as also would be the reflecting efficiency of the mirror itself. The area needed by our family of four rises therefore to a monstrous total of 1200 m². Remembering that the mirror must be steerable to follow the Sun, the required system would be like a fair-sized radiotelescope. Indeed the big radiotelescope at Jodrell Bank (Figure 9.3) has an area of 4500 m², and would accordingly suffice only for four

Figure 9.3 The Jodrell Bank radiotelescope. The house on the right gives an idea of scale.

families. It would be necessary to provide a structure like the Jodrell Bank telescope for even a very small cluster of houses.

The incremental steps from an initial area of 20 m² to the very much larger 1200 m², and the reasons for the steps, are set out in Table 9.1. The exceedingly poor overall efficiency of the system is clearly indicative of a wrong engineering concept. Nor would the situation be changed by replacing the solar cell by a water boiler at the mirror focus, because the efficiency of conversion of steam to electricity is not much different from 20 per cent.

Table 9.1 *Increments to an initial area of 20 m²*

Area (m²)	Reason for increment
100	20 per cent efficiency of solar cell.
300	Ratio of day to night (winter).
600	Blocking of sunlight by cloud, and absorption by atmospheric gases and particles.
1200	Losses in electricity storage and in reflection at mirror surface.

Unless solar power stations in space can be developed in the long term (21st century) there is no satisfactory high-technology method for adapting 'solar' to the more sophisticated needs of modern society. The problem is not of energy *quantity* but of energy *concentration*. To be industrially effective, an energy source must be compact, which is precisely where nuclear wins and solar fails.

The concept of energy concentration is many centuries old. Figure 9.4 illustrates an ancient technology for the deployment of solar power, a technology excellent by modern standards, a technology that works by the subtle courtesy of biology. Solar energy is first stored over many years in the wood of trees. After the trees are cut, wood from an area is collected into piles—the first stage of concentration. The wood is then smouldered into charcoal, so driving off its more capricious, less-controllable aromatic contents. The resulting charcoal is now burned in a final stage driven by powerful bellows, which concentrate the flame into the fiercest possible heat. All this was understood by craftsmen many centuries ago. If it were understood today by sociologists, media personnel, and many others who feel their views on energy should be heard, less nonsense would be talked and written.

It is a well-known fact of history that England's ancient forests had become seriously denuded already in the 16th century, when the English population was only 5 million instead of the present 50 million. If the present *world* population were 500 million instead of 4000 million, Figure 9.4 would still represent a satisfactory technology.

Another good way to make use of solar power is through the fermentation of vegetable material to alcohol, a process that has been known for many thousands of years. Every major oil company has long ago costed

Figure 9.4 The trompe used to blow a forge fire. Note the tilt-hammer in the foreground. 1678. (From *A History of Technology*, Volume III, Oxford, 1958)

the industrial production of alcohol as a means for replacing natural petroleum fuels by artificially-produced fuel. In 1975 dollars, the cost works out at about $30 per barrel of oil equivalent, about twice the present cost of actual oil. The extra $15 per barrel would result in an excess cost of about 25p per gallon of fuel. Since most Europeans already pay some $2 or more per gallon of motor fuel, the increase occasioned by a switch from OPEC oil to industrial alcohol or to wood alcohol would hardly cause a social revolution.

Soft energy paths

We mention here a book* under this title by A. B. Lovins. The book contains rather little about 'soft' paths. It is mostly an attack upon 'hard' paths, which is to say upon oil, coal, and nuclear. The format consists largely of quotations and graphs taken from public documents, arranged in a kind of patchwork. The links between the patches are not always easy to follow.

Our concern with the book is that it has become a formal reference for the anti-nuclear movement. Members of this movement, when confronted by a request for how they themselves would provide energy, fend off the issue with the remark: 'It is all in Lovins.' It is not all in Lovins. Possibly we may have missed others, but the only original calculation we found in the book is on page 44, and this calculation contains an eleven hundred per cent error:

> 'Each year the U.S. beer and wine industry, for example, microbiologically produces 5 per cent as many gallons (not all alcohol, of course) as the U.S. oil industry produces gasoline. Thus a conversion industry roughly ten to fourteen times the physical scale (in gallons of fluid output per year) of U.S. cellars and breweries, albeit using different processes, would produce roughly one-third of the present gasoline requirements of the United States.'

In spite of the disclaimer, this calculation assumes wine and beer to be entirely pure alcohol. The actual situation for the year 1971 is shown in Table 9.2†

Table 9.2 *Per capita consumption levels (litres) for the United States*

Category	Consumption by volume	Consumption by alcohol content
Spirits	9.92	4.47
Wine	7.87	1.25
Beer	98.03	4.43

Adding the two columns one sees that, while the total per capita volume consumption in 1971 was 115.82 litres, the alcohol content of the consumed drinks was, thankfully, only 10.15 litres. The ratio of total volume to alcohol content was therefore 11.41, the factor by which Lovins' calculation is wrong. Of course even the most careful author may be overtaken by an oversight, or even by a real mistake. What is exceptional about this case, and why we mention it, is that it was allowed to stand.

Alcohol production would therefore have to be increased by 300, not by 30 as Lovins claims, to equal U.S. gasoline consumption. Would such a very large increase really be possible one might wonder? The answer is

* Penguin Books, Harmondsworth, 1977
† *Encyclopaedia Brittanica 1975*, Volume 1, p. 444

probably yes, but not easily so, not by vegetable matter grown on fertile land, unless food production was to be seriously impeded. The growth would need to come from marginal lands, and perhaps from the great forests of the tropics.

Yet liquid fuel for cars is only one aspect of our energy need. There is no prospect of alcohol production replacing the major part of present-day energy sources. Quite apart from the limited availability of vegetable material, there is a fundamental issue of principle still to be considered. It is essential for a primary energy source that it shall *deliver more energy than it consumes*, a condition that is by no means obviously satisfied for alcohol production. Energy is needed to grow and to harvest the necessary vegetable matter, to collect it together and to process it into alcohol. In a society with access to natural oil at $15 per barrel, the economic cost of $30 per barrel equivalent of industrial alcohol (or of wood alcohol—methanol) itself implies a considerable energy expenditure.* One can suspect a rather close-run situation in which alcohol production consumes pretty well as much energy as it delivers. So what would be the point of it? The point would be convenience. Granted ample primary energy from *other sources*, which might be in the form of electricity, there would be the convenience of using the alcohol to run vehicles and to fly planes.

Solar power thus fails again as a primary energy source, even in one of its most attractive applications. In a society with abundant energy from nuclear sources, however, alcohol production could work well as a secondary energy source. There is no reason why in such a society vehicles should not continue to run and airplanes to fly as they do at present.

In spite of the unattractive situation for 'solar', there is no shortage of politicians ready to espouse its cause, both in Europe and the United States. And so long as contracts are being awarded for its development there will be no shortage of commercial companies, large and small, anxious to get into solar. The attraction to politicians lies in the vote-gathering possibilities. If only solar could somehow be made to work efficiently, thus confounding the opinion of technical experts through the world, a politician could represent himself to the people as the St. George who slew the nuclear dragon. Sooner or later, however, there will of course be a day of serious reckoning, as breakdowns of the electricity supply become more frequent, and as shortages develop of petrol for the family car and of heating for the home. When this day comes such politicians will be forced into early retirement, but unfortunately the damage they are causing will live on after them.

---

* Calculations at 1975 prices.

10
Natural gas

Natural gas is the energy source that can probably be increased in its supply most rapidly on a short-term basis. But if orthodox geology is correct in its explanation of the origin of natural gas as a biological phenomenon, there is no prospect of it contributing in an important way to the world's energy requirements much beyond the last years of this century. On such a view gas will become exhausted along with oil, and at roughly the same rate. So there would be little reason for giving an extended discussion of natural gas here if it were not for another possibility recently suggested by Professor T. Gold, who thinks most natural gas has been mixed with interior rocks since the origin of the Earth itself. According to Gold, natural gas is mostly of mineral origin and is rising all the time to the surface from considerable depths within the Earth. In this connection it is interesting to recall from Chapter 3 a remark in the quotation from *Harper's Weekly*:

> 'It is one of the tenets of the geologist that the lower we go for coal the more gas we strike.'

Gold advances a considerable body of evidence in favour of his theory. He argues that natural gas emerges in mud-volcanoes, which sometimes occur in regions where there is no expectation of biogenic deposits being present. Moreover, gas samples from mud-volcanoes indicate a rather pure hydrocarbon, whereas gas samples from wells of biogenic origin tend to be contaminated with a complex of carbon compounds. Mud-volcanoes are found north of the Caucasus, in Burma, in the Caribbean, in S. America, and in central Australia. A mud-volcano that flared up recently near Baku in the Soviet Union is said to have produced a flame 2 km high which burned for eight hours, consuming perhaps a million tons of natural gas.

Gold also argues strongly for a connection between the emergence of natural gas from the Earth and the occurrence of earthquakes. He points out that eyewitness accounts of large earthquakes in antiquity contain remarks such as 'the sky was alight' or 'flames shot out of the ground', which could have been escaping gas subjected to spark ignition. In the past few years, claims to predict earthquakes from a curious medley of phenomena have been made. These include a clustering of small shocks before the main tremor, an increase in the concentration of radioactive radon gas above the earthquake zone, rises and falls of level in water wells accompanied by abnormal numbers of bubbles, changes in the electrical properties of the ground itself, the uplift of considerable areas of land, and reports of strange animal behaviour. Gold seeks to unify these apparently diverse symptoms of an earthquake in terms of the escape of natural gas in large quantities and at great pressure from the Earth's interior. The gas makes an upward journey to the surface by way of the fractures caused by the earthquake, pushing with it other gases, such as radon, which are also contained within the Earth. Near the surface it displaces carbon dioxide and volatile agents associated with fungi and other plants. It is these 'earthy' smells emerging from the Earth which produce the strange behaviour of animals equipped with keener noses than ourselves. In the aftermath of earthquakes many dead fish have been reported at the sea surface, a phenomenon according to Gold caused by sulphurous gases coming up from the bottom of the ocean.

The theory also explains the cause of tsunamis, the giant waves that sweep from time to time over low-lying areas of oceanic islands. These waves sometimes extend over a front many hundreds of miles long. They are caused in Gold's view by gigantic bubbles of gas that emerge from the ocean floor, and by a large-scale upwelling of water as the vast bubbles rise and break up when they approach the sea surface. It is this raising of the water which then causes the tsunami water wall on the surface of the sea.

If this theory is correct, and we think there is a good chance that it is, there is the prospect of being able to recover from the deeper regions of the Earth's crust large and previously unsuspected supplies of natural gas. There is evidence of this happening already in Pennsylvania, where companies prospecting for gas and oil ran into what at first seemed like a number of dry wells. Faced with the choice of cutting their losses or of continuing to drill still deeper, the prospectors chose to continue down below the level where any biological material might have been expected. They struck a rich reservoir of almost pure methane gas, the lightest of the hydrocarbons. When these drillings were plotted on a map it was found that the most successful wells lay along an old and deep geological fault, which could well be providing egress from the lower crust. How much the supply of natural gas can be increased by drilling in this way along fault lines remains to be investigated.

A second more ambitious way to increase the supply of natural gas would be to drill still more deeply into the Earth's crust, so as to provide artificial exit shafts for the gas. That is to say, to take out the gas at a faster than normal rate. In the long run according to the theory such deep exit shafts could stop earthquakes, but in the short and medium term it might well cause them, and as Gold himself points out this is not a possibility to be taken lightly.

Earthquakes are destroyers of human life on a far greater scale than any of the disasters we have considered to this point. An earthquake at Agadir, Morocco, killed 12 000 in 1960. The worst earthquake on record, which occurred on 23 January, 1556 in the Shensi Province of China, is said to have killed 800 000. Disasters on this scale are almost impossible for the mind to grasp, and paradoxically we are more apt to ignore them than to emphasise them. It is to be doubted that many in Europe or America were very strongly impressed by the so-called 'circular storm' of 12–13 November, 1970, in which 1 000 000 people of the Ganges Delta Islands are said to have lost their lives.

Suppose that Professor Gold's theory is correct, and suppose that with deep-drilling it becomes possible to acquire a supply of natural gas adequate to obviate the need for nuclear reactors. At what level of risk of tsunamis and earthquakes would we abandon nuclear for natural gas? With this question in mind, we end the present chapter with still another quotation from Welman and Jackson's *Disaster Illustrated* (*Ibid* p. 14):

'It was early evening, and downtown Anchorage was filled with people leaving work. It was Good Friday (1964) and everybody was looking forward to the Easter weekend. But they hadn't gotten very far when, without warning, the streets began to ripple and pitch beneath them. People tried to scramble to safety, but there was nowhere to go. Thirty blocks of downtown Anchorage were crumbling. Crevasses 12 feet deep and 50 feet wide opened in the streets, and cars, people, and buildings slipped away. There was nothing to do but watch helplessly, and to try to brace against the relentless shocks. The same scene and worse was being enacted in the ring of towns surrounding Anchorage.

'At Seward, Kodiak, Whittier, and Valdez—coastal towns —seismic waves did more damage than the quake. At Valdez, a hole opened up in the dock area and a man and two children were swallowed up in it. Moments later, a pier with 12 stevedores on it disappeared. The scenes at Valdez were so horrifying that its residents moved out afterward, never to return.

When it was over, the Alaskans learned that 131 people had died. The shock, centered at Port Arthur, some 75 miles distant from Anchorage, was one of the strongest ever recorded in the United States. It far exceeded the shocks that destroyed San Francisco. Only the thin distribution of the population and the lack of high skyscrapers had kept the fatalities down.

11
Coal

Unless it proves possible to increase accessible reserves of natural gas sharply and dramatically along the unorthodox lines discussed in the preceding chapter, coal will remain by far the greatest non-nuclear energy resource. But coal has always been an inconvenient material. It has always been dangerous to win out of the ground, as we saw in Chapter 3, and it has always been a source of pollution. In Volume III of *A History of Technology*,* J. U. Nef writes:

> Already during the early decades of the 17th century coal came into widespread use, not only in the domestic hearths of the English and Scottish, and in their laundry-work and cooking, but in the extraction of salt and the manufacture of glass, bricks and tiles for building, anchors for ships, and tobacco-pipes. The dyers, the hat-makers, the sugar-refiners, the brewers, who were growing very numerous especially in London and some provincial towns, and even some of the bakers of bread required coal. . . . . In the year 1563–64 the shipments of coal from Newcastle-upon-Tyne amounted to 32,951 tons. A century later, in the year 1658–9, they had risen to 529,032 tons. Between about 1580, when Shakespeare is said to have settled in the capital, and the Restoration in 1660, the imports at London increased some twenty- to twenty-five-fold. Foreigners who visited the rapidly growing city were astonished at the filthy smoke from tens of thousands of domestic fires and from hundreds of workshops. With its breweries, its soap- and starch-houses, its brick-kilns, sugar-refineries, earthenware works, and glass-furnaces, London seemed to some of these foreigners to have been rendered unfit for human habitation. Even the English virtuoso John Evelyn (1620–1706) was repelled by the fog of smoke belching from the sooty throats of the new manufacturing shops, to hang over the metropolis and insinuate itself along the streets. He compared this new, dark London to 'the picture of Troy sacked by the Greeks . . . .'

Later in this chapter we shall find a somewhat similar situation still persisting. Yet it was coal that saved civilization as the demands of rising populations denuded the forests of Europe:

* Oxford University Press, 1958, page 76.

The future of an expanding manufacture of iron in the British Isles after about 1600 came to be bound up with the replacement of wood by coal in the furnaces and forges at which pig iron was converted into bar iron, some of which was converted into rods at slitting-mills.

The discovery which eventually lifted the use of coal from merely a noxious source of heat, usable only in the crudest of circumstances, lay in its conversion to coke. To quote Professor Nef yet again:

'It was in the drying of malt, necessary for certain brews, that the new fuel transmitted its obnoxious properties indirectly to the taste of the beer. Few persons could bear to drink beer brewed from malt dried with raw coal. The idea of charring coal, as wood was charred to produce charcoal, to purge the mineral fuel of some of its impurities, may have occurred in 1603 to an ingenious promoter named Sir Hugh Platt (1552–1608), who supplied a recipe for making briquettes as a means of sweetening the domestic fires that caused so much distaste to sensitive noses in London. But the early efforts to coke the coal failed. It was apparently in connexion with the drying of malt that success was first achieved, in Derbyshire about the time of the Civil War (1642–8). Beer brewed from malt dried with what were then called 'coaks' was pronounced sweet and pure, and, as a result of the new discovery, Derbyshire beer became famous throughout England.'

With the development of modern science, it has been realised that coal is a rich source of organic chemicals, and that its greater value should come from these chemicals, instead of coal being largely used as a source of heat. Ironically, it is just these important chemicals that are driven off and lost in the production of coke. When eventually nuclear energy has replaced coal as a primary energy source, it will then be as a source of chemical materials that coal will at last be properly used and valued.

It seems, however, that before this day comes, coal will once again be called on to bale out society as a primary energy source, as oil supplies go over the peak from about 1985 onwards. The length of time for which coal will thus be 'wasted' will depend on the extent to which nuclear energy is delayed and impeded over the coming decade.

Much work is going on at the moment, with a view to resolving age-old coal pollution problems. It needs to be done, because it seems that the much-trumpeted solutions of a few years ago are no solutions at all, and may indeed only have made matters worse than they were before. Thus T. Alexander writes in the issue of *Fortune* for November 20, 1978:

NEW FEARS SURROUND THE SHIFT TO COAL

'Recently . . . a number of scientific findings have been coalescing to suggest that the environmental cost of any large-scale reversion to a coal-based economy could be dangerously high [see Figure 11.1]. In fact, some of the measures initiated in the past to reduce the damage have actually made things worse. Though these issues have not escaped environmental leaders and policy-makers, they have been

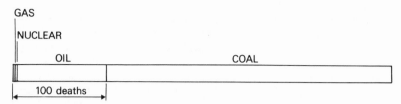

Figure 11.1 The human cost of generating electric power using various fuels. These represent the highest estimates of fatalities that a single 1000 megawatt plant of each type could cause annually due to occupational exposures, environmental effects, and accidents. Estimates were gathered by the American Medical Association.

discussed in uncharacteristically muted tones, usually out of public earshot. Perhaps this is because environmentalists are defensive about some apparently misguided air-pollution policies they promoted in the past, and fearful that any derogation of coal will promote nuclear energy, the alternative many of them dislike even more. Professor Richard Wilson, director of Harvard University's Center for Energy and Environmental Policy, refers to the situation as a "conspiracy to whitewash coal".

'Coal production and consumption have always been burdened with an assortment of ancillary costs. Underground mining, for instance, has traditionally exacted more casualties per man-hour than any other major occupation. Along with its harmful effects on human health, coal mining has always imposed heavy aesthetic and ecological penalties. The notched profiles and flattened peaks of large sections of central Appalachia will probably rank among mankind's most enduring marks upon the planet. Another blight that is likely to be with us into perpetuity is the acid drainage from abandoned mines. This has rendered some 10000 miles of Appalachian streams biologically sterile. No satisfactory cure for acid mine-drainage seems to exist. The main remedy is to dose the mines or streams periodically with neutralizing lime, but this creates messy calcium-sulfate sludges in the stream beds that are hardly more wholesome than the acid.

'For society at large, coal's most noxious and obnoxious drawback has always been its smoke. Of all the fuels that man employs, coal produces the most copious and complex array of emissions thought to be damaging to health. But how these emissions actually do their damage has long been a matter of guesswork. It now appears that our control policies were based on the wrong guess.

'For hundreds of years, most of the blame for the noxiousness of coal combustion has focused on the sulfur that is present in all coal in varying amounts and that combines with oxygen during combustion to form sulfur dioxide (SO_2). In the late Fifties, epidemiologists came to suspect SO_2 as the principal culprit in what they referred to as the "urban factor"—the long-observed fact that mortality rates tend to be higher in cities than in the country. The most impressive evidence appeared when meteorological temperature inversions held pollutants trapped low over coal-burning industrial regions of cities for days at a time, leading to hundreds or thousands of "excess" hospitalizations and deaths from respiratory ailments.

'Still, scientists have found puzzling anomalies in the evidence against sulfur dioxide. For example, laboratory workers who induced animals or human subjects to breathe SO_2 at concentrations even higher than those in polluted cities often found that the subjects didn't seem particularly distressed. The conventional explanation for these results until recently was that in order for SO_2 to show harmful effects, it had to be breathed in conjunction with particulate matter such as soot or fly ash. Scientists assumed that the SO_2 attached itself to the particles, which then lodged in sensitive tissues of the respiratory passages and lungs. They also believed that much of the damage was caused by the oxides of nitrogen (NOX) and hydro-carbons emitted by factories, power plants, and automobiles.

'Therefore, the control strategy that cities, states, and eventually the federal government employed was to set hypothetical thresholds, or "no effects" levels, below which the pollutants might be regarded as harmless. With science largely mute as to what such thresholds might be, the permissible levels were determined by the political process.

'A variety of tactics were adopted to meet those levels. They included stiff emission controls on automobiles, which are abundant producers of NOX, hydrocarbons, and carbon monoxide. Factories and power plants were required to switch to low-sulfur fuels—oil, gas, or special coals—particularly during unfavorable weather conditions. Coal boilers were equipped with electrostatic precipitators that could remove more than 90 per cent of the soot and fly ash, and pollution-emitting new plants were situated far from cities and fitted with tall smokestacks so that the pollutants could be diluted before they reached the ground.

'Among the gratifying consequences of these measures was that by 1975, the tonnage of large particulates in the atmosphere of most large cities had declined remarkably, and SO_2 had declined somewhat. Yet some scientists continued to puzzle over contradictory evidence about SO_2, evidence that raised doubts about whether a threshold level could actually be said to exist and even whether the SO_2/particulate complex was especially harmful. The evidence included findings from some long and painstaking experiments at the Harvard School of Public Health by toxicologist Mary O. Amdur, who is now at M.I.T. Since 1949, Amdur had been exposing laboratory animals not only to SO_2 itself but also to a variety of sulfate compounds that are slowly formed as SO_2 mixes with oxygen and trace substances commonly found in polluted atmospheres. One of these sulfates, for instance, is ordinary sulfuric acid. Sulfates usually emerge in the form of ultrafine droplets or infinitesimal solid particles that tend to remain suspended in the atmosphere. Amdur found that several sulfates—principally sulfuric acid, but also sulfates of zinc and iron—were many times more irritating to the respiratory tracts of guinea pigs than oxides of nitrogen, hydrocarbons, and SO_2 plus particles. From the work of Amdur and others, most experts have concluded that most of the damage in urban air pollution comes from sulfates.

'Obviously, the evidence about sulfates does not alter the practical necessity to control the SO_2 precursor. But it has powerful implications for what the control tactics should be. It means, for instance, that sulfur pollution is *not* a local problem, as was long assumed. Rather, since the minute sulfate particles can be carried by winds for thousands of miles, they are long-range agents that affect dispersed populations. As Harvard's Richard Wilson points out, the tall-stack/electrostatic-precipitator remedies may well have the effect of *worsening* the hazard: when SO_2 is emitted close to the ground, it tends to adhere to leaves and other surfaces. The tall stacks provide the time for the SO_2 to convert to sulfates, which can be carried far and wide by winds. All this helped to explain a puzzle that pollution monitors were already pondering: while SO_2 levels had been diminishing within the cities, the sulfate levels had often been *increasing*, not only in cities in the eastern U.S. but in the countryside as well. The prevailing winds in the U.S., flow from west to east, so the northeastern states tend to be on the receiving end of sulfates produced elsewhere.

'More obvious than their effects upon human health has been the apparent impact of the sulfates upon the ecology of certain regions. A strange silence reigns in the environmental movement about this threat, which seems far more serious than the logging depredations or menaces to snail darters that have brought the movement lunging into court. For as sulfates—as well as nitrates, which also get formed from pollutants—are wafted across the country on westerly winds, they eventually become incorporated into precipitation as dilute mixtures of sulfuric and nitric acid. Some of the rainwater gathered in New England in recent years measures several hundred times more acid than normal rainwater, or roughly the equivalent of vinegar.

'This acid rain can have interesting effects on the earth it falls upon. Most of what is known about these effects comes from Scandinavia, which for decades has received acid blown in from all over industrial Europe. Because high acidity inhibits many aquatic species and the bacteria that cause decay, Sweden's lakes are becoming empty of fish and all but a few choking species of matlike sphagnum moss that can tolerate high acidity. Foresters have also discovered that the growth rate of Sweden's largely coniferous forests has been declining since the 1920s, a phenomenon they attribute to acid rain and snow.

'Only in this decade have U.S. scientists found that many lakes in the northeastern U.S. are in the same general shape. Many of the lakes in New York's Adirondacks, which had abundant trout in the 1930s, are devoid of these fish now. And some northeastern forests have apparently been stunted in the same way as Sweden's.'

It is likely that the above situation, described so effectively by Mr. Alexander, will turn out more an indictment of environmentalist and so-called 'consumerist' groups, and of the regulatory agencies with which Washington D.C. is swarming like a house infested by cockroaches, than it will be an eventual indictment of coal itself. A letter by A. Wormser in the December 18 edition of *Fortune* made a number of relevant points in this respect:

'Fluid bed combustion, one of the near-term new technologies, appears to be so efficient at reducing air pollutants that coal may actually become a clean fuel like natural gas. To be sure, the fluid beds now being introduced are only satisfactory: while meeting pollution standards, they are not ultraclean. But under some circumstances, fbc has reduced both SO_2 and NOX emissions to barely detectable levels. Visible smoke, the most obviously objectionable pollutant of traditional coal systems, is now so efficiently cleaned by baghouses that chimneys of coal systems are significantly cleaner than those of systems burning oil.'

There is an overwhelming argument, not mentioned by Mr. Alexander, against a long-term reliance on coal. Coal is not distributed uniformly over the world, as will be seen from Figure 11.2. While both the U.S.S.R. and the U.S.A. might contemplate such a long-term reliance, the rest of the industrialised world cannot do so—at any rate without becoming energy-dependent on the U.S.S.R. and the U.S.A. Such a sensitive dependence would be a certain recipe for political unrest.

It is true that world resources of nuclear fuel are not uniformly distributed either, but they are better distributed than coal. And by using what are called breeder reactors, which we shall discuss in Chapter 14, the need for more than small quantities of nuclear fuel can eventually be avoided. With breeders, it is hard to see that any sovereign state could be barred from access to uranium, especially as uranium could be extracted if need be from seawater, which with very few exceptions is available to everyone.

This brings us to the close of four chapters in which we have considered energy sources other than nuclear. In the next chapter we turn to an issue currently being argued by environmentalists against nuclear energy, the issue of waste disposal, which we shall find to be no issue at all. Much more relevent are the dangers of non-nuclear technologies, dangers that we have emphasised in former chapters, and which are partially displayed in Figure 11.1. In concluding the present chapter we take a look at these dangers in the quantified form recently given by Dr.

Table 11.1 *Risk in man-days lost per unit energy output*

	Upper limit of estimates	Lower limit of estimates
Coal	2500	110
Oil	1800	12
Nuclear	10	1·7
Natural gas	6	–
Ocean thermal	30	25
Wind	700	125
Solar		
Space heating	125	100
Thermal	600	80
Photovoltaic	700	160
Methanol	350	225

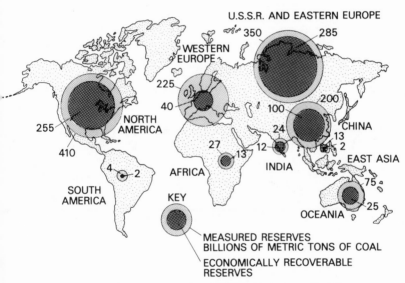

Figure 11.2 Total world reserves of coal. 'Measured reserves' are in well-surveyed deposits whose extent and quality have been established by adequate sampling. Most of the reserves are located in the Northern Hemisphere, but the rest of the world is likely to increase its share through new exploration. (Adapted from E. D. Griffith and A. W. Clarke, *Scientific American*, Volume 140, 1979.)

H. Inhaber,* set out in Table 11.1. Man-days lost are through deaths, injuries, and disease, and the energy unit is the megawatt-year (1 megawatt-year = 8.76 million kWh). Methanol is wood alcohol—the 'soft' technology favoured by A. B. Lovins (Chapter 9). Concerning methanol, Dr. Inhaber writes:

* *Science*. Vol. 203, 718, 1979.

'The remarkably high occupational risk for methanol is primarily due to one factor—logging. In Canada (and elsewhere in the world), this is a job with quite high accident rates.'

The high values for wind and solar are perhaps unexpected, and Dr. Inhaber remarks:

'What are the reasons for these surprising rankings? The details are contained in a recent (AECB) report.* The main reason why non-conventional systems have relatively high risk is the large amount of materials and labour they require per unit energy output. Why should solar need more materials than coal or oil? It's because of the diffuse nature of the incoming energy: solar and wind energy are weak, and require large collection and storage systems to amass an appreciable quantity of energy. Coal, oil, and nuclear systems deal with concentrated forms of energy and so require less apparatus. This argument is simplistic and glosses over many lesser considerations, but is generally found to be true.

'The large quantity of materials required for unconventional systems implies huge industrial efforts in mining, fabricating, and constructing the collectors, storage systems and all related apparatus. Every form of industrial activity has an associated risk, which can be found through accident statistics compiled by national organisations. When all the multiplications and additions are done, we find that the risk from unconventional energy systems can be substantial.

'This raises an interesting point. Although these systems are labelled unconventional, their risk comes, in the main, from highly conventional sources. In other words, the risk from windmills doesn't come primarily from a blade flying off and hitting you on the head, and the risk from solar space heating doesn't arise from falling off the roof as you make that last little adjustment. Rather, it comes from the more mundane tasks of mining the coal, iron and other raw materials and fabricating them into steel, copper and glass.'

---

* *Risk of Energy Production*. 1978, No. AECB-1119. Atomic Energy Control Board, PO Box 1046, Ottawa, Canada, K1P 5S9.

12

Storing one's own nuclear waste

It is easy for the anti-nuclear activist to make up stories about the dangers of the waste products of the nuclear industry, because if one wishes to do so it is easy to make up stories about the dangers of any aspect of technology. To be hit by a stream of molten steel would be instantly fatal. Society does not deal with this problem by disbanding the steel industry, however, but by concentrating steel-making plants in such a way that the public at large does not come into contact with molten iron. In a like fashion, the nuclear industry concentrates its reactors, and the waste products from the reactors, so that the public is not exposed to serious radioactive hazards.

Waste products can be turned into a glassy material that does not leach easily in contact with water. The vitreous material can be shaped into cylinders about a foot in diameter and ten feet long. Enclosed in strong metal casings, the cylinders can then be buried below ground at depths of 3000 feet or more, when for all practical purposes they could just as well be 3000 light years away from us.

If Britain derived all its electricity from nuclear fission, the total amount of waste produced each year would amount to only about a hundred such cylinders. This small amount may be compared with the vastly greater quantities of waste generated by other industries, which other waste is simply left lying around in surface tips or dumped into rivers and into the sea.

There is an atavistic superstition, however, which helps the anti-nuclear activists to spread the notion that something darkly horrible will happen if nuclear waste is buried underground. We all tend to be repelled by the concept of an 'underworld'. Since time immemorial, the dead have been buried underground. Hades lay below ground in Greek mythology, and modern preachers are still given to pointing downward in making their references to Hell. We can almost bring ourselves to feel that burying nuclear waste deep at 3000 feet would be worse than burying it shallow—it would be nearer to Hell—whereas deep burial is objectively better than shallow burial.

It is also easy to bring oneself to believe that, however carefully the storage site were chosen, sooner or later water would come into contact with the cylinders of nuclear waste. The metal casings of the cylinders would gradually corrode and be eventually penetrated by the water, which would proceed to leach the vitreous radioactive waste. Water with radioactive materials dissolved in it would then rise upward to emerge at ground level, it is claimed, where the materials would become incorporated in the soil, so passing eventually into the food of humans and of animals generally. How, it may be asked, can one be certain over periods of many centuries that this disaster scenario will not come about?

Places with little or no water do exist, and could be drilled and excavated for storage sites, if residents in such places permitted it. Whenever geologists seek to drill in an appropriate area, however, anti-nuclear activists make a sustained effort to alarm local residents, with the avowed purpose of interfering with the best method for the disposal of wastes. It seems that if certain of the anti-nuclear critics could find a way to provoke a nuclear accident they would be only too happy to cause it.

Even if an underground storage site is now quite dry, can one be sure that it will remain dry for several millennia into the future, by which time the radioactivity of nuclear waste would have declined a millionfold or more? Probably yes, because geological changes in stable rock formations occur much more slowly than this. Yet consistent with the 'worst case' point of view of the nuclear industry, investigations have been made on the assumption that waste becomes exposed to water action. The American Physical Society report mentioned in Chapter 5 makes the extreme concession of dispensing with all protection for the waste—the steel casing of the cylinders and the vitrification are supposed absent. Thus the waste is taken in the study to go immediately into water solution. Calculations extending 800 000 years into the future were made. After even this great span of time, nothing had percolated upward from a depth of 3000 feet to the surface (*ibid*, page S126).

One might think that, in the face of these arguments, the protesters would now desist. They do not desist, however, because their aim is not to make correct scientific arguments but to make emotional propaganda. There is almost no end to the imaginary scenarios which a deliberately obstructive critic can invent, and responsible people who understand that an energy crisis is fast approaching must at some point lose patience.

Very well, argued Bernard L. Cohen in the issue of *Scientific American* for June 1977, suppose if you will that the nuclear industry is infested by monsters whose aim is to use nuclear waste to harm as many people as possible. How would their activities compare with those of fellow-monsters in the chemical industry? To avoid making things too easy for the chemical monsters, Professor Cohen allowed them only materials that are handled routinely on an everyday basis—making up special poisons, like the dioxin which escaped at Seveso, N. Italy, was not permitted. Even so, of all the possibilities considered, chemical and nuclear, the biggest potential killer turned out to be chlorine, the stuff we use so freely to clean baths and swimming pools. Administered in concentrated doses, chlorine would be 1000 times more devastating to human life than nuclear waste (also in deliberately concentrated doses). Phosgene would be 100 times more devastating. Prussic acid and simple ammonia would each be 10 times more devastating, while the compounds of barium (scarcely known to the public) would be about the same as nuclear waste.

While this competition between rival monsters in the chemical and nuclear industries was of course fictional, Professor Cohen pointed out that, whereas the nuclear industry proposes to bury its waste deep in the ground, far out of harm's way, potential chemical poisons remain at ground level. No normal person is proposing to eat or breathe nuclear materials deliberately. On the other hand dangerous chemicals are sprayed routinely onto crops that we really do propose to eat deliberately.

Furthermore, some of the chemical poisons, the compounds of arsenic and barium for example, do not disappear. They are with us for eternity. Nuclear waste, on the other hand, declines steeply in its activity as the years pass—after 300 years the reduction is to as little as $\frac{1}{100\,000}$ of the initial activity.

Although these arguments would appear to be fully sufficient to answer the critics, it will be a help towards getting an instinctive feeling for the scale of the waste-disposal problem if we rediscuss the whole matter from a more personal point of view. Suppose we are required individually to be responsible for the long-term storage of all the waste that we ourselves, our families and our forebears, have generated in an all-nuclear energy economy.

It will be useful to think of waste in terms of the categories of Table 12.1.

Table 12.1   *Categories of nuclear waste and their lifetimes*

	Lifetime (years)
High-level	10
Medium-level	300
Low-level	100 000
Very low-level	10 million

High-level waste is carefully stored over its 10-year lifetime by the nuclear industry. This is done above-ground in sealed tanks. It is not proposed to bury nuclear waste underground until activity has fallen to the medium-level category of Table 12.1. Instead of underground burial, however, we now consider that medium-level waste is delivered for safe keeping to individual households.

We take the amount of the waste so delivered to be that which has been generated over the 70 years from 1990 to 2060. If by the mid-21st century the promises of those who favour so-called soft technologies have borne fruit, or if solar-power satellites have proved economically viable, of if there really are vast stores of methane waiting to be tapped inside the Earth, society can dispense with nuclear power should it wish to do so.

But if a satisfactory alternative to nuclear fission has not emerged by the mid-21st century, then nuclear fission will be essential, and its use will necessarily be non-controversial at that stage.

The controversial aspects of nuclear fission are therefore concerned only with the period 1990 to 2060, and over this period a typical family of four would accumulate $4 \times 70 = 280$ person-years of vitrified nuclear waste, which for an all-nuclear energy economy would weigh about 2 kilograms. Supplied inside a thick metal case, capable of withstanding a house fire or a flood, the waste would form an object of about the size of a small orange, which it could be made to resemble in colour and surface texture—this would ensure that any superficial damage to the object could easily be noticed and immediately rectified by the nuclear industry.

The radioactive materials inside the orange would be in no danger of getting smeared around the house, not like jam or honey. The radioactive materials would stay put inside the metal orange-skin. Indeed the orange would be safe to handle freely but for the γ-rays emerging from it all the time. The effect on a person of the γ-rays would be like the X-rays used by the medical profession. If one were to stand for a minute at a distance of about 5 yards from the newly-acquired orange, the radiation dose received would be comparable to a medical X-ray.

Unlike particles of matter, γ-rays do not stay around. Once emitted, γ-rays exist only for a fleeting moment, during which brief time they are absorbed and destroyed by the material through which they pass. Some readers will be familiar with the massive stone walls of old houses and barns in the north of England. If a γ-ray emitting orange were placed behind a well-made stone wall 2 feet thick, one could lounge in safety for days on the shielded side, and for a wall 3 feet thick one would be safe for a lifetime.

Our family of four would therefore build a small thick-walled cubicle inside the home to ensure safe storage of the family orange.

After several generations, the waste inside the orange would have declined to the low-level category of Table 12.1, when the orange could be taken out of its cubicle and safely admired for an hour or two as a family heirloom. It could also be returned to the nuclear industry for reprocessing into a considerably smaller object of the size of a gull's egg, and the family could specify which particular species of gull it should be made to resemble.

We could go into details on the construction of the cubicle, how it could be arranged to conduct out the small amount of heat generated by the nuclear waste, how it could be reinforced to withstand earthquakes as well as fire and flood, and how it could be monitored for safety by external controls. But manifestly the problem for our family is not primarily one of safety. The problem would be boredom, the boredom of storing a gull's egg down many generations.

Such individual tedium would of course be avoided if the waste were stored communally. For 100 000 families making up a town of 400 000 people there would be 100 000 eggs to store. Or since it would surely be inconvenient to maintain a watch on so many objects the town would have the eggs reprocessed into a few hundred larger objects of the size of pumpkins or vegetable marrows. The whole lot could be fitted into a garden-produce shed, except that instead of a wooden wall, the shed would need to have thick walls of stone or metal.

This then is the full extent of the nuclear-waste problem that our own generation is called on to face. If by the mid-21st century it has become clear that nuclear fission is the only effective long-term source of energy, society will then have to consider the problem of accumulating waste on a longer time-scale. For the town of 400 000 people, a shed of pumpkins would accumulate for each 70 years, until the oldest waste fell at last into the very low-level category of Table 12.1, when it could be discarded.* After 7000 years, there would be a hundred sheds, which could be put together to make a moderate-sized warehouse. In 100 000 years there would be about 15 medium warehouses, which could be accumulated into two or three large warehouses. Thereafter, the problem would remain always the same, with the oldest waste falling into the very low-level category as fast as new waste was generated. Of course, the 'warehouses' would be deep underground, as we discussed earlier in this chapter, and there would be no contact between them and the population of the town.

It is to be doubted, however, that we should worry ourselves too much about the far-distant future. Our problem now is to get through just the next generation or two with an on-going new energy source coming into full operation as fossil fuels decline. Unless this problem is solved, there will be no far-distant future worth worrying about.

The nuclear-waste problem, if not exactly trivial, is plainly not a cause for serious concern. The risk that each of us would incur, even if called upon to store our own waste, would be insignificant compared with the risks we routinely incur in other aspects of our daily lives.

---

* Weight for weight, very low-level waste has about the same activity as uranium ore. Since uranium ore is handled in annual quantities that run into millions of tons, on an energy-equivalent basis with much greater safety than coal-mining, the disposal of very low-level waste would not be a problem.

13

A nuclear accident in the U.S.S.R.?

A report on a Soviet nuclear disaster occurring in the U.S.S.R. has frequently been quoted as a case example to show that nuclear energy is unsafe. Whereas in the West, it is argued, disasters have been prevented just in the nick of time, in the U.S.S.R. a disaster really happened. The disaster has sometimes even been represented as an explosion of a nuclear reactor, seemingly giving credence to the notion that reactors can explode.

This report has been dismissed by nuclear experts in the West, because an explosion involving radioactive materials thrown up into the air would have been detected by equipment flown in western intelligence aeroplanes, if not indeed by the cows at Windscale. An explosion of sufficient violence to throw material through a horizontal distance of 30 kilometres, the claimed extent of the disaster area, would also throw material high into the upper atmosphere, from which it would take some years to fall-out, and from which it would become spread by upper atmosphere movements, spread far outside Soviet air space.

To sustain the claim of an explosion, it would therefore be necessary to suppose a cover-up of information by western intelligence services. While such an insinuation would be in keeping with the emotive state of the arguments, commonsense suggests otherwise. Western intelligence did not cover up the test explosions of nuclear devices in the U.S.S.R., or in China, or later in India. Indeed the third bar in Figure 5.1, the bar for nuclear explosion fall-out, has been calculated from precisely this kind of information, which is publicly available.

Nor at the date in question, 1957–58, was there any sensitive issue involving the civil nuclear industry to be covered up. In 1957–58, plenty of oil was available, and nuclear energy was not then seen as essential, and there was no public uproar over it. The uproar has started only after need has become urgent.

This was the situation at the time of writing *Energy or Extinction*, when one of us took the official position described above. At about that time, Dr. Zhores Medvedev, the author of the report in question, published a more detailed account in the *New Scientist* (30 June 1977), which began:

'In my article "Two decades of dissidence" (*New Scientist*, vol 72, p. 264), I mentioned the occurrence at the end of 1957 or beginning of 1958 of a nuclear disaster in the southern Urals. I described how the disaster had resulted from a sudden explosion involving nuclear waste stored in underground shelters, not far from where the first Soviet military reactors had been built; how strong winds carried a mixture of radioactive products and soil over a large area, probably more than a thousand square miles in size; and how many villages and small towns were not evacuated on time, probably causing the deaths later of several hundred people from radiation sickness.

'I was unaware at the time that this nuclear disaster was absolutely unknown to Western experts, and my *New Scientist* article created an unexpected sensation. Reports about this 20-year-old nuclear disaster appeared in almost all the major newspapers. At the same time, some Western nuclear experts, including the chairman of the United Kingdom Atomic Energy Authority, Sir John Hill, tried to dismiss my story as "science-fiction", "rubbish" or a "figment of the imagination".

'However, about a month later my story was confirmed by Professor Lev Tumerman, former head of the biophysics laboratory at the Institute of Molecular Biology in Moscow, who had emigrated to Israel in 1972. Tumerman visited the area between the two Ural cities—Cheliabinsk and Sverdlovsk—in 1960. He was able to see that hundreds of square miles of land there had been so heavily contaminated by radioactive wastes that the area was forbidden territory. All the villages and small towns had been destroyed so as to make the dangerous zone uninhabitable and to prevent the evacuated people from returning. Tumerman's eye-witness evidence did not, however, convince all the experts, including Sir John Hill, of the truth of this disaster. Doubts remained that the story was exaggerated. These doubts convinced me of the need to collect more information that would throw light on the real scale of this nuclear disaster.'

Dr. Medvedev goes on to discuss a paper by F. Rovinsky published in 1966 (*Atomnaya Energiya*, Volume 18, p 379). Two highly radioactive lakes are discussed in the region slightly east of a line from Sverdlovsk to Chelyabinsk, 'the first was 11.3 sq. km. in size and the second was 4.5 sq. km., both round in shape'. A third highly contaminated lake is described in two papers published by A. I. Il'enko (*Voprosy Ichtiologii*, Volume 10, p 1127, and Volume 12, p 174).

More recently still, the *New Scientist* (1 Dec. 1977) carried a news item under the headline

CIA CONFIRMS MEDVEDEV'S DISASTER CLAIM

'The CIA released the documents in response to a Freedom of Information Act request from the Critical Mass Energy Project, set up by Ralph Nader to chronicle the experience of operating nuclear plants. The project received 14 heavily censored documents from the CIA, which at the same time withheld another 15, allegedly too sensitive to be released.

'The information provided in the CIA documents is mainly anecdotal. There are vivid accounts of the aftermath of at least two nuclear accidents in the Kyshtym area of the southern Urals. The first accident allegedly occurred in the spring of 1958, and the second was sometime in 1960 or 1961. One report, dated March 25, 1961, quoted an unnamed source as saying "hundreds of people perished and the area became and will remain radioactive for many years". The source said he had visited the "strange, uninhabited and unfarmed area . . . Highway signs along the way warned drivers not to stop for the next 20 to 30 kilometres because of radiation. The land was empty. There were no villages, no towns, no people, no cultivated land. Only the chimneys of destroyed houses remained."'

In view of the evidence of this CIA report, and of the papers cited by Dr. Medvedev, it must be accepted that a nuclear incident of some kind occurred about 20 years ago in the region 150 km to the south-east of Sverdlovsk. We also accept Dr. Medvedev's argument that three lakes would not be deliberately contaminated by radioactivity merely to serve as research laboratories. On the other hand, three small lakes might well have been deliberately contaminated if it had been thought by the Soviet government that such an action would have the effect of supplying the anti-nuclear movement in the West with sufficient propaganda to halt Western civil nuclear developments, and so making the West eventually dependent on Soviet coal (Figure 11.2).

However, one must doubt whether the Soviets would believe sufficiently in the ineptness of Western governments to foul their own nest deliberately in this way. So we must look towards a natural and rational explanation of the key facts, which are:

(1) A nuclear incident of some kind occurred in the Sverdlovsk area at some time around 1958,
(2) The first Soviet military reactors had been sited in that area,
(3) There was no explosion in the usual meaning of the word, because otherwise upper-atmosphere radioactivity would have been detected by Western intelligence.

Let us look for a sensible scenario consistent with these three points. During the Second World War the Ukraine and much of Russia proper was overrun by German armies. More than 20 million people are said to have lost their lives. In such a situation it is hardly credible that the U.S.S.R. could have sustained much of a research programme in nuclear physics, certainly nothing to compare with the Manhattan Project in the United States. The first explosions of nuclear bombs in 1945 by the U.S. must therefore have come as a profound shock to the Soviet government, not because the Soviets had been unaware of the distant possibility of making such bombs, but because the Americans had managed to go so far so quickly.

As a matter of the most urgent priority, Moscow must then have ordered the inception of a programme for producing nuclear weapons, to begin at the earliest moment (in late 1945 or 1946). The first steps in this programme must inevitably have proceeded on a corner-cutting basis, with risks being taken that would have been avoided in a less acutely-pressing strategic situation. Inevitably too, the quickest route to nuclear weapon production would be seen to lie, not in the construction of a huge gas-diffusion plant for separating the isotopes of uranium, but by building a military reactor (point 2, above) for producing plutonium.

Waste from such a reactor would eventually have to be treated, or disposed of in some way. In appreciable quantity, such waste is hot, in the literal sense of producing much heat. Normally hot waste is placed in liquid contained in a carefully sealed tank. The heat is transferred to the metal wall of the tank, from the outer surface of which it is removed harmlessly, either by simply radiating the heat, or by an external flow of cooling water passing over the wall of the tank.

If, however, in the early Soviet situation, such a safe method of cooling waste were not available, it is conceivable that somebody had the superficially plausible idea of sealing the waste in metal drums and of then dumping the drums in two or three small lakes. The metal walls of the drums would prevent the waste going into the water—so it would be argued—while the lake water would serve to keep the drums cool. For workers operating under severe technical difficulties, this might well have seemed a simple solution to the problem of waste disposal.

The time would have been around 1947. Ten years elapsed, and the drums began to leak, perhaps when attempts were made to raise them to the surface for safer relocation. By 1957, the lake waters had thus become highly radioactive, as Dr. Medvedev claims. All that remained then was for gale-force winds to lift radioactive spray from the lakes over the surrounding countryside.

We do not claim this to be the only rational explanation of the facts, but we do suspect that Dr. Medvedev's report is to be explained in some such way, and we think that Dr. Medvedev might have arrived at such a straightforward explanation for himself. We claim, moreover, that it makes no sense to stir up misgivings about an incident over which the full circumstances have not been published. It would surely be a great irony if the Soviets, through making mistakes themselves, were to succeed in deferring the civil nuclear programme of the West.

14

Breeder reactors and the U.S. Non-proliferation Act

It has been an important tenet of economics that every commodity is freely available at a price determined by the relation of supply to demand, an assumption that has not always been true. The supply of whale oil, used in the 19th century for candles and lamps, could not be increased indefinitely, no matter the price offered for it.

In such exceptional cases it has usually been found that a steeply rising cost for one commodity encourages the development of alternatives capable of doing the same job—natural oil replaced whale oil, and if the price of a particular metal goes too high, some other material usually takes over its function.

From 1973, however, this replacement concept has not operated at all well for energy commodities, and this has been one of the reasons why orthodox economic theories have been at sixes-and-sevens over the past few years. The sudden quadrupling of the price of oil soon drove up the prices of both coal and of uranium, the nuclear fuel. The latter rose per pound weight (0.454 kg) from a pre-1973 range of \$5–10 to the present \$30–45.

Uranium is crucial to the nuclear industry because it is the only naturally occurring material to contain 'fissile' atoms, namely atoms of uranium-235. A fissile atom is one that can be made to undergo fission in a simple convenient way involving what are known as 'slow' reactors, of which the pressurised light water reactor mentioned below is an example. Most of uranium consists of atoms of the non-fissile isotope uranium-238. About 99.28 per cent of natural uranium is uranium-238 and only 0.72 per cent is uranium-235.

World resources of high-grade ores containing 0.1 per cent or more of uranium oxide do not amount in contained uranium to more than about 5 million tons, but dropping the ore grade to the range 0.0025 to 0.025 per cent increases the amount of uranium available many times. For sufficiently low grades, there is an essentially unlimited quantity of uranium in the rocks of the Earth's crust, so that, unlike coal and oil, *there is no problem of a complete exhaustion of the uranium supply*.

It might seem therefore as if the economists' dictum, that there will always be plenty provided the price is right, would apply to uranium. There is, however, an overriding physical dictum:

A primary energy supply is not viable, if access to the relevant energy commodity consumes more energy than the supply itself delivers.

This constraint will negate solar-power satellites, certainly over the near and intermediate terms, and possibly even in the long term. This constraint is indeed a cloud hanging over most solar-energy devices, which nearly always consume more energy in their manufacture than they deliver in their use. It is also a constraint to be applied to the availability of uranium. More energy must not be used to dig and crush a low-grade ore, and to separate out the uranium, than nuclear technology obtains from the uranium itself.

The pressurised light water reactor, the poor man's technology

The least efficient nuclear technology is that of the pressurised light water reactor, operating without reprocessing to extract the plutonium-239 produced from uranium-238 within the reactor. The burn-up fraction of uranium is then only about 0.5 per cent. At this poor efficiency, 1 pound weight (0.454 kg) of uranium yields only about 15 000 kilowatt-hours of electricity,* worth in money terms about $400. Remembering that we have to count the energy required to build the reactor, the energy required to distribute the electricity to consumers, the energy needed to keep employees of the electricity and mining companies alive (and to supply them with cars, radios, etc.), it is clear that not more than about a half of the money-value of the 15 000 kWh can be diverted to the extraction of 1 pound of uranium, not more than about $200. At this price, not more than about 10 million tons of uranium is available the world over, and at the very poor burn-up efficiency of only 0.5 per cent the energy yield from 10 million tons of uranium is not sufficient to meet the world's total electricity requirement for more than a few decades.

Shortly after coming to office, President Carter caused the U.S. Non-proliferation Act to be passed by Congress. The Act sought to impose the poor technology described in the previous paragraph, not only on the United States, but on the rest of the democratic world, thereby handing a major technological advantage to the Warsaw Pact nations whose economies were not affected by the Act, which sought to impose itself through economic sanctions, working for example in the following way. A German company supplying the home nuclear industry with computers could be prevented from importing silicon chips from the U.S., unless the German nuclear industry complied with the terms of the Act. This pressure on the internal affairs of other nations explains the recent urgency of the British Government in supporting the indigenous manufacture of silicon chips.

* Taking the efficiency of heat to electricity conversion to be 30 per cent.

The pressurised light-water reactor works with what is known as enriched uranium, which is to say uranium from which about three-quarters of the non-fissile U-238 has been removed, thereby raising the concentration of the fissile U-235 to about 3 per cent of the U-238. The older technology for enriching uranium was cumbersome and difficult, and until recently the U.S. had essentially a monopoly throughout the non-communist world in its production. The pressurised light-water reactor being of American design, in order to encourage its export, and to persuade other nations to build it under licence, the U.S. had agreed to supply enriched uranium, subject only to certain quite liberal conditions being satisfied. It was these liberal conditions which the new Non-proliferation Act replaced by the stringent conditions described above, even though the original agreements had not yet run their course.

The French government immediately protested that such a cancellation was illegal, which it was, if international agreements are to be worth anything at all, even between friends. Indeed the French seem to have decided to wash their hands of such agreements and to go it alone. Moving full steam ahead with the most advanced of all nuclear technologies, that of the fast breeder reactor, the French now seem destined to forge ahead of the rest of Europe and of the United States. Events and policies can of course change quickly, but if present showing is maintained France must emerge in the 21st century as a dominant industrial power.

The fast breeder, the rich man's technology

The non-fissile uranium-238, constituting more than 99 per cent of natural uranium, is converted in a fast-breeder to fissile plutonium. For 50 per cent efficiency this increases by 100 the energy yield from one pound weight of uranium, from the previous 15 000 kWh of electricity to 1.5M kWh, worth about $40 000. Not only does one pound weight of uranium produce 100 times more electricity than before, but we can now afford far more energy to dig up and refine the very large quantity of uranium that is present in low-grade ores. Indeed we can afford the energy-equivalent of many thousands of dollars of money. The economists' doctrine of ample commodity supply, provided the price is right, now become entirely valid. The quantity of uranium available at tens of thousands of dollars per pound weight is essentially unlimited, giving an unlimited energy economy for the future.

This situation was known to European leaders at their meeting in London with President Carter at the time of the passing of the U.S. Non-proliferation Act. The point of view of the French we have noted already. British policy was to make encouraging noises without information content, but to follow quietly behind the French, a policy that can lead to no great harm so long as the French backs continue to run the ball. British policy was reminiscent of a story which Lord Miles tells about the late Archbishop Ramsey. Observing that in greeting a substantial number of persons, one following another, the Archbishop was perpetually making a mellifluous cooing but incoherent sound, Bernard Miles asked Ramsey why he did so. The Archbishop's reply was: 'So that people cannot engage me in conversation.'

The most sorely tried of the European leaders was Chancellor Helmut Schmidt of W. Germany. German resources of the higher grades of coal are indifferent, and German reserves of oil and gas are negligible. Nuclear energy is therefore a more critical matter for Germany than it is for any other major industrialised country except Japan. Precisely because of this urgency, the anti-nuclear movement is particularly strong in Germany, and President Carter's new policy 'initiative' played strongly into the hands of the German anti-nuclear movement. Chancellor Schmidt also had his own left-wing political party to reckon with. The U.S.S.R., like France, is firmly pressing ahead with its nuclear programme. One might therefore expect left-wing parties in the democracies to be in favour of nuclear energy, but it is in the nature of the peculiar patriotism of the far left-wing to support nuclear energy in the U.S.S.R., but to disapprove of it in their own countries.

Middle-class technologies

Nevertheless it seems to have been Chancellor Schmidt who caused President Carter to make a slight shift from his commitment to the poor man's technology. The shift consisted in the setting up of an international study group with the code letters INFCE, standing for International Nuclear Fuel Cycle Evaluation. This study is concerned with the many ways that uranium (and thorium) can be used to deliver energy, with particular reference to ways intermediate between the extremes discussed above. The aim of INFCE is to find, if possible, a nuclear technology that satisfies the following conditions:

(1) the system is required to be economically superior to the poor man's technology,

(2) the system is required to prevent the proliferation of nuclear weapons.

Although INFCE is not scheduled to report until 1980, it is already rumoured that (2) cannot be met, in which case the fast breeder would not necessarily be worse in respect of proliferation than intermediate technologies. In *Energy or Extinction*, one of us favoured an intermediate fuel cycle involving both uranium and thorium, in the hope of being able to satisfy (2). However, because of improvements (outside the nuclear industry) of isotope separation techniques, which were mentioned briefly in *Energy or Extinction* and which will be discussed here in Chapter 15, former hopes in this respect for the uranium-thorium cycle may not be realisable. Indeed, as we shall also discuss in Chapter 15, the U.S. policy of supplying the world with enriched uranium looks as though it is likely to be a greater source of weapons proliferation than the plutonium-239 produced by fast breeders.

Since the publication of *Energy or Extinction*, we have been impressed by the number of nuclear engineers who have written in support of the fast breeder. The feasibility of the uranium-thorium cycle has not been challenged, but one correspondent likened it to a model-T Ford, a sound, reliable job, but then added: 'Why build a model-T when you can build a Rolls-Royce?'

Physicists outside the nuclear industry tend to think of the fast breeder as a difficult technical proposition. Inside the nuclear industry however, it appears to be regarded as a thoroughly feasible proposition. Probably those nearest to the actual work have the sounder judgment of what is often referred to as 'the state of the art'. We are inclined therefore to take the nuclear industry at its word, and to believe that a fast breeder reactor, giving very high fuel efficiency, and therefore an unlimited energy future, really can be built without serious difficulties.

15

Who is doing the proliferating?

In our hearing, an anti-nuclear campaigner remarked to a reporter on one of the London newspapers: 'Plutonium proliferation is the big thing with us now.' A big thing too with President Carter and the U.S. Non-proliferation Act. In this chapter we shall consider the concept of plutonium proliferation.

Weapons proliferation means that more nations than before come to possess nuclear bombs. By the year 1970, five nations had tested some form of nuclear weapon: in chronological order, the U.S.A., the U.S.S.R., Britain, France, and China. Since 1970, India has also tested a nuclear explosive.

The concept of proliferation in the sense of the above quotation means something rather different, however. The concern is that less-advanced nations might find a way to obtain nuclear weapons through civil energy-producing reactors and facilities supplied to them by technically more advanced nations, as for instance W. Germany currently holds a contract to supply civil energy-producing reactors and facilities to Brazil. The anti-proliferation argument is that such contracts as that between W. Germany and Brazil should be avoided (or broken if they exist already) except when drawn up under the restricted and very energy-inefficient conditions demanded by the U.S. Non-proliferation Act.

It is hard to see, however, why any nation wishing to acquire a nuclear weapon should start from a civil energy-producing reactor. In fact none ever has, for the good reason that another much surer, far less publicly-exposed way, of achieving the same end is readily available. To understand this other way, we note that India acquired a nuclear explosive by building a so-called research reactor, not by ordering a commercial reactor from Europe or America. It is true that both Canada and the United States gave help, particularly through a supply of heavy water from Canada. But if heavy water had not been supplied, Indian scientists could have opted for another of the many possible reactor designs. The production of a nuclear explosive would then have taken the Indian Government a little longer, but it could not have been stopped.

Given time and $100 million (some authors set the cost much lower) any sovereign state can acquire a number of plutonium bombs. The necessary primary supply of uranium need not be bought on the world market. The rocks of the Earth's crust contain uranium in varying concentrations. High-grade ores with about 0.25 per cent uranium are rather unusual, occurring mainly in the U.S., Canada, and Australia. But rocks with a uranium content of 0.025 per cent are to be found in almost every country, and from such rocks uranium can still be extracted. The penalty for using low-grade ores is cost, and if cost is no object the supply of uranium is everywhere for the taking.

Next, the design and construction of a research reactor and of a bomb-making facility needs about a hundred physicists, engineers, and chemists. On this point, the recent report of the American Physical Society (*ibid*, page S29) has the following to say:

> 'In spite of classification of weapons technology and design concepts, information has gradually become available not only to groups of experts but also to individual members of the public. As we review in more detail in Chapter VI of this report, the design principles for fission explosives are distributed quite widely in the open literature. A good example is provided by the Encyclopedia Americana article by John S. Foster, who is a well-known expert on nuclear weapon technology and formerly the Director of one of the AEC's (Atomic Energy Commission) weapons development laboratories. His article presents a broad view of the nature of the nuclear explosion and requirements for its initiation. In addition, skilled people can use other information now in the public domain that was originally classified . . .
>
> 'A wide range of information is published in technical literature concerning the chemistry and metallurgy of plutonium and uranium. Of the two metals, uranium is the easier to handle by far . . . However, the procedures for handling plutonium are also well described. The required chemical and metallurgical apparatus for small scale operation also is available on the open market . . .'

The reader may well be aghast at this situation, and may even feel that it only reinforces the need to stop all future nuclear developments. Such a view is emotive, however. All that can be stopped now is the good side of nuclear energy, the energy-producing side; the bad side is already out of the bag. While one can regret that this is so, it has to be remembered that at no stage of human history has it ever been possible to suppress the spread of technical discoveries, a truth already well-known to the writer of the Book of Genesis.

Even if a sovereign state is initially without technically qualified people, the only problem is time. All that needs to be done then is to send fifty to a hundred intelligent students to the United States for training at the excellent laboratories to be found in American universities. Masquerading under the façade of education, the training would doubtless receive the enthusiastic support of the U.S. Government. Our information is that students from third-world countries are indeed being so trained at the present time.

If energetically pursued, such a weapons project would take 10 to 15 years from inception to bomb-testing. Although the existence of the project would probably become known to intelligence agencies in the developed countries, there would be nothing of the glare of publicity in the world's media that the misuse of civil facilities (made available under international treaty) would be certain to provoke. The world has long accepted the convention that nations which make their own nuclear bombs are substantially freed from criticism, otherwise the high and mighty (including those responsible for the U.S. Non-proliferation Act) would be the first to be convicted.

The U.S. Non-proliferation Act was the hasty work of a newly-elected President, whose advisers were seemingly unaware of the revolution of isotope separation technology that had been taking place quietly over the previous decade. Although a full assessment of this question will have to await the publication in 1980 of the report of INFCE (International Nuclear Fuel Cycle Evaluation) many scientists think that because of developments in such techniques the U.S. Non-proliferation Act may well turn out to be a step towards, not away from, weapons proliferation.

The general nature of an atomic explosion is well-known. A neutron entering the nucleus of a heavy atom causes the nucleus to split up into two more or less comparable pieces. The pieces then emit further neutrons which enter the nuclei of other heavy atoms, provoking further splitting and further neutron production in a cycle that amplifies itself provided the quantity of the heavy atoms is large enough.

To date, uranium-235 and plutonium-239 have been the kinds of heavy atom used, and of these uranium-235 is much less difficult to work with as a bomb-making material. But in the past uranium-235 has been far harder to obtain than plutonium-239, because uranium-235 constitutes only 0.72 per cent of natural uranium, and separating such a small fraction was formerly an exceedingly cumbersome business, beyond the capability of all but the most technically advanced nations. Modern ultra high-speed centrifuges are in the process of changing this situation, however, which means that, in the event of a nation really seeking to acquire nuclear bombs from a commercial energy-producing reactor, the inevitable question must arise: Is the separation of uranium-235 from the fuel cells of a reactor not a better proposition nowadays for bomb-making than the older more awkward use of plutonium? Sooner or later, with the improvement of isotope separation techniques, a stage must come when an affirmative answer is given to this question. In the opinion of some scientists that stage is now.

With this perspective, let us recall the terms of the U.S. Non-proliferation Act. For those nations agreeing not to obtain additional energy from plutonium-239, and so accepting the least efficient fuel cycle of Chapter 14, supplies of enriched uranium are to be made available. For nations not so agreeing, supplies of even natural uranium are to be withdrawn, all technical assistance is to be refused, and various forms of economic sanctions are to be imposed as well (Chapter 14).

Enriched uranium is uranium from which most of the common uranium-238 has been removed, usually about three-quarters of it, so that the concentration of uranium-235 is raised from 0.72 per cent to about 3 per cent. Supplying enriched uranium therefore solves the major part of the problem of obtaining weapons-grade uranium-235. A nation that built itself a centrifuge to obtain weapons-grade uranium-235 would thus find that the United States, by supplying the enriched uranium, had assisted the weapons proliferation, to the extent of removing three-quarters of the uranium-238.

Turning now from sovereign states to subnational groups, fortunately there is little prospect of dissident groups being able to build, or to acquire, a reactor of their own. Discussions of weapons-proliferation by terrorists have therefore been concerned with the possibility of such groups abstracting bomb-making materials clandestinely from national facilities. On this question we quote again from the report of the American Physical Society (page S94):

> 'First, we make the assumption that the technology for isotope separation is not available to *subnational* groups ...
>
> 'Table VI-1 (*here Table 15.1*) summarizes the nuclear safeguards concepts which have been proposed ... *Deterrence* indicates the ability of a safeguards measure to discourage would-be thieves; *containment* indicates the ability of a safeguards measure to reduce the probability that an attempted theft will succeed; the *recoverability* indicates effectiveness for preventing misuse of stolen special nuclear materials (SNM) either by leading to recovery of the material itself or by degrading the materials with respect to unauthorized use.'

The reader interested in the detailed significance of the entries in Table 15.1 will find information and further references in the American Physical Society report. We are not concerned here so much with details, however, as to emphasise that the nuclear industry has been well ahead of its critics in the consideration of these matters. Just as the nuclear industry has worked for almost two decades within the constraints of a 'maximum credible accident', so it now plans to prevent the 'maximum credible theft' of special nuclear materials.

The precautions listed in Table 15.1 would be quite insurmountable for an outside group of terrorists. The technical problems would defeat anything other than an inside job, and for this there is the entry 'Personnel considerations' in the Table. There have been objections to 'Personnel considerations', on the grounds that surveillance and vetting of personnel would constitute an infringement of the liberties of workers in the nuclear industry. Perhaps the objectors have overlooked the fact that similar vetting is applied in banks, and that there is already strict surveillance of gold, diamonds, and indeed of all valuable commodities. Nobody is required to work in the nuclear industry, or in a bank, and

Table 15.1 *Proposed nuclear safeguards concepts, grouped according to safeguards function and according to nature of barrier presented to theft and subsequent use of SNM.*

	Deterrence	*Containment*	*Recoverability*
Physical security barriers	Specially designed containers. Guard forces. Surveillance, alarms. Special communications. Personnel considerations. Co-location of fuel cycle facilities.	Specially designed containers. Guard forces. Surveillance, alarms. Special communications.	Specially designed containers. Homing devices.
Technical barriers	Materials accountability. Restricted fuel forms. Dilution. Protective radiation hazard.	Dilution Protective radiation hazard.	Degraded SNM Chemically released tracers. Dilution. Protective radiation hazard.

those who are highly sensitive about their liberties are perfectly free to work elsewhere. More likely, the objectors are irritated at the care which has already been taken to avoid trouble in a situation where they had hoped to find it.

This chapter has been concerned with deliberately induced nuclear explosions, not with the possibilities of inadvertent accidents that we discussed in earlier chapters. A nuclear explosion releases more energy than a single conventional weapon, and it is of course the sudden release of energy which constitutes the violence of an explosion. This consideration suggests that the frightfulness of the weapons possessed by society depends predominantly on the amount of their energy release, not on the source of the energy. The explosion of a nuclear weapon would therefore produce the same measure of frightfulness as a sudden release of many chemical bombs totalling to the same amount of energy. The conventional air-raid on Dresden released an amount of energy comparable to the nuclear raid on Hiroshima, which explains why the numbers of victims in the two attacks were similar, 135 000 at Dresden, 90 000 at Hiroshima.

Unless a society commands significantly more explosive nuclear energy than chemical energy, it follows that nuclear weapons do not add greatly to the net frightfulness of the society. In the current so-called peacetime, it is probable that the nuclear explosive energy available to the 'super-powers' is appreciably greater than their arsenals of chemical explosives, but if we think back to the First and Second World Wars a considerable fraction of the whole productivity of the combatants was then absorbed in the making of chemical explosives. We are therefore led to the rather surprising (and probably quite unpopular) conclusion that nuclear weapons have not added to the potential explosive violence of society by anything like the big factor that is usually supposed. The largest effect has come from the fusion energy in so-called hydrogen bombs, rather than from fission bombs of the Hiroshima type.

The popular view is dominated by the fact that the packets of energy released by nuclear weapons are very much bigger than the packets released by chemical weapons. The latter are much more numerous, however, of the order of 100 000 times more numerous. The larger packets in which nuclear energy is released have altered the strategy of war. The situation in the First World War, in which young men were sent away to fight under appalling conditions, while governments and civilian populations continued to sit in comparative comfort at home, has gone for ever. Nuclear weapons can reach civilian populations and governments. Indeed, governments are now more of a target than the infantryman in the field, and without a doubt this fact has contributed to the prevention of a third world war.

Because they involve such large energy packets, nuclear weapons are inefficient against the individual. For a similar energy output, nuclear weapons could never have had the efficiency of the tiny contact-explosive chemical devices which the Russians scattered by the million 'in the woods of the Smolensk area, and which are said to have killed more than half a million individuals of the German army during the winter of 1941–42.

For similar reasons, nuclear bombs are very probably far too complex and too large in their scale to be suited to the activities of terrorist groups, particularly bombs of plutonium-239, about which the anti-nuclear activists are so worried. The making of an effective plutonium bomb is by no means the weekend's exercise for an intelligent schoolboy that it is often said to be. Highly-skilled scientists took more than a year to make such a bomb, and even up to the actual test-explosion of the first plutonium bomb in 1945 the effectiveness of the proposed design remained in doubt.

According to headlines in the media one might think that society is overflowing with people anxious to commit crimes of violence. Yet the situation is exactly the reverse, otherwise civilisation would for long have been impossible. When one considers the enormous quantity of potentially explosive material to which we all have ready access, and which causes the many inadvertent accidents that we have mentioned in previous chapters, the number of crimes of violence is astonishingly few. It is vastly more difficult to make a plutonium bomb than it is to make a chemical explosive from petrol. Yet each of us has easy access every year to about a ton of petrol, so that for Britain as a whole there is sufficient material within easy access to produce an annual total of some tens of megatons of explosive chemicals, the equivalent of several hundred plutonium bombs.

16

It has all happened before

Although the technologies are of course very different, the situation today for the nuclear industry is psychologically very similar to the situation in 1825 at the beginning of the railway age. The pressing need then for efficient communication between Manchester and Liverpool was widely agreed and was not in question, just as the pressing need for a new energy source is now agreed and is not in question. There were then vociferous groups recommending 'soft' technologies, as for instance barges drawn by horses on canals, groups that like the anti-nuclear movement were very litigious, groups that hated the hard technology of the railway, and were prepared to go to any lengths to suppress it.

Just as today the opponents of nuclear energy scorn the high competence of the nuclear industry, so the opponents of the railway scorned the competence of the engineer chosen by the Liverpool and Manchester Railway Company, a colliery mechanic named George Stephenson. Just as a number of distinguished scientists can be found today inveighing against the nuclear industry, so George Stephenson was faced by engineers and scientists whose high reputations were belied by the nonsense they talked, nonsense about alternatives to the railway and the locomotive, alternatives that read uncannily like the current nonsense of power from the Sun, and from wind and water.

In 1825 a self-educated mechanic took on himself the animosity of the anti-railway activists. It is interesting that the Committee of the House of Commons, before which George Stephenson was called, showed no perception of the revolution that was to be initiated by the railway age. Not only did the colliery mechanic understand the technology of the railway itself, but he also had a clearer idea of the social changes to come than did those who set themselves up to legislate for the nation. This circumstance too has its modern parallels.

The following quotations are from the classic biographies of George Stephenson and of his son Robert, by Samuel Smiles.*

'On the 25th of April, 1825 George Stephenson was called into the witness-box. It was his first appearance before a Committee of the

* *The Lives of Engineers* (David & Charles), 1968.

House of Commons, and he well knew what he had to expect. He stood before the Committee to prove what the public opinion of that day held to be impossible. Clear though the subject was to himself, and familiar as he was with the powers of the locomotive, it was no easy task for him to bring home his convictions, or even to convey his meaning, to the less informed minds of his hearers. In his strong Northumbrian dialect, he struggled for utterance, in the face of the sneers, interruptions, and ridicule of the opponents of the measure, and even of the Committee, some of whom shook their heads and whispered doubts as to his sanity, when he energetically avowed that he could make the locomotive go at the rate of 12 miles an hour!

'And yet his large experience of railways and locomotives, as described by himself to the Committee, entitled (him) to speak with confidence on such a subject. Beginning with his experience as a brakesman at Killingworth in 1803, he went on to state that he was appointed to take the entire charge of the steam-engines in 1813, and had superintended the railroads connected with numerous collieries from that time downwards. He had constructed fifty-five steam-engines, of which sixteen were locomotives. The engines constructed by him for the working of the Killingworth Railroad, eleven years before, had continued steadily at work ever since, and fulfilled his most sanguine expectations. He was prepared to prove the safety of working high-pressure locomotives on a railroad, and the superiority of this mode of transporting goods over all others. As to the charge that locomotives on a railroad would so terrify the horses in the neighbourhood, that to travel on horseback or to plough the adjoining fields would be rendered highly dangerous, the witness said that horses learnt to take no notice of them. In the neighbourhood of Killingworth, the cattle in the fields went on grazing while the engines passed them, and the farmers made no complaints.

'Mr. Alderson, who had carefully studied the subject, and was well skilled in practical science, subjected the witness to a protracted and severe cross-examination as to the speed and power of the locomotive, the stroke of the piston, the slipping of the wheels upon the rails, and various other points of detail. Mr. Stephenson insisted that no slipping took place, as attempted to be extorted from him by the counsel. He said, "It is impossible for slipping to take place so long as the adhesive weight of the wheel upon the rail is greater than the weight to be dragged after it." As to accidents, Stephenson said he knew of none that had occurred with his engines. There had been one, he was told, at the Middleton Colliery, near Leeds, with a Blenkinsop engine. The driver had been in liquor, and put a considerable load on the safety-valve, so that upon going forward the engine blew up and the man was killed. But he added, if proper precautions had been used with that boiler, the accident could not have happened.'

We pause to note a similarity to the recent emergency at Three Mile Island, Pennsylvania, although at Harrisburg not a man was killed or injured.

'The Committee also seem to have entertained considerable alarm as to the high rate of speed which had been spoken of, and proceeded to examine the witness further on the subject. They supposed the case of the engine being upset when going at 9 miles an hour, and asked what, in such a case, would become of the cargo astern. To which the witness replied that it would not be upset. One of the members of the Committee pressed the witness a little further. He put the following case: "Suppose, now, one of these engines to be going along a railroad at the rate of 9 or 10 miles an hour, and that a cow were to stray upon the line and get in the way of the engine, would not that, think you, be a very awkward circumstance?" "Yes," replied the witness, with a twinkle in his eye, "very awkward—*for the coo!*" The honourable member did not proceed further with his cross-examination. Another asked if animals would not be very much frightened by the engine passing them, especially by the glare of the red-hot chimney? "But how would they know that it wasn't painted?" said the witness.'

'The case of the opponents was next gone into, in the course of which the counsel indulged in strong vituperation against the witnesses for the Bill. One of them spoke of the utter impossibility of making a railway upon so treacherous a material as Chat Moss, which was declared to be an immense mass of pulp, and nothing else. "It actually," said Mr. Harrison, "rises in height, from the rain swelling it like a sponge, and sinks again in dry weather; and if a boring instrument is put into it, it sinks immediately by its own weight. The making of an embankment out of this pulpy, wet moss, is no very easy task. Who but Mr. Stephenson would have thought of entering into Chat Moss, carrying it out almost like wet dung? It is ignorance almost inconceivable. It is perfect madness, in a person called upon to speak on a scientific subject, to propose such a plan. Every part of this scheme shows that this man has applied himself to a subject of which he has no knowledge, and to which he has no science to apply."

'Evidence was given at great length showing the utter impossibility of forming a road of any kind upon Chat Moss. Mr. Francis Giles, C.E., had been twenty-two years an engineer and could speak with some authority. "*No engineer in his senses,*" said he, "would go through Chat Moss if he wanted to make a railroad from Liverpool to Manchester. In my judgement *a railroad certainly cannot be safely made over Chat Moss without going to the bottom of the Moss*. The soil ought all to be taken out, undoubtedly; in doing which, it will not be practicable to approach each end of the cutting, as you make it, with the carriages. No carriages would stand upon the Moss short of the bottom. My estimate for the whole cutting and embankment over Chat Moss is £270 000, nearly, at those quantities and those prices which are decidedly correct. It will be necessary to take this Moss completely out at the bottom, in order to make a solid road." '

What George Stephenson had seen, although he seems to have kept the idea to himself at the House of Commons Committee, probably because of fear of ridicule, was that by building his railroad on a raft to spread the weight the problem could be solved. And so it eventually was:

'The road across Chat Moss was finished by 1 January 1830, when the first experimental train of passengers passed over it, drawn by the "Rocket"; and it turned out that, instead of being the most expensive part of the line, it was about the cheapest. The total cost of forming the line over the Moss was £28 000, whereas Mr. Giles's estimate was £270 000! It also proved to be one of the best portions of the railway. Being a floating road, it was smooth and easy to run upon, just as Dr. Arnott's water-bed is soft and easy to lie upon—the pressure being equal at all points. There was, and still is, a sort of springiness in the road over the Moss, such as is felt in passing along a suspended bridge; and those who looked along the line as a train passed over it, said they could observe a waviness, such as precedes and follows a skater upon ice.

'During the progress of these works the most ridiculous rumours were set afloat. The drivers of stage-coaches brought the alarming intelligence into Manchester from time to time, that "Chat Moss was blown up!" "Hundreds of men and horses had sunk, and the works were completely abandoned!" (Mr. Stephenson) himself was declared to have been swallowed up in the bog, and "railways were at an end for ever!" '

Because it was a *railroad* that was being constructed, nowadays we are apt to think of a locomotive drawing carriages upon the rails. But nothing of the sort had been decided:

'The works of the Liverpool and Manchester Railway were now approaching completion. But, singular to say, the directors had not yet decided as to the tractive power to be employed in working the line when opened for traffic. The differences of opinion among them were so great as apparently to be irreconcilable. It was necessary, however, that they should come to some decision without further loss of time, and many Board meetings were accordingly held to discuss the subject. The old-fashioned and well-tried system of horse haulage was not without its advocates; but, looking at the large amount of traffic which there was to be conveyed, and at the probable delay in the transit from station to station if this method were adopted, the directors, after a visit made by them to the Northumberland and Durham railways in 1828, came to the conclusion that the employment of horse-power was inadmissible.

'Fixed engines had many advocates; the locomotive very few: it stood as yet almost in a minority of one—George Stephenson. The prejudice against the employment of the latter power had even increased since the Liverpool and Manchester Bill underwent its first ordeal in the House of Commons. In proof of this, we may mention that the Newcastle and Carlisle Railway Act was conceded in 1829, on the express condition that it should *not* be worked by locomotives, but by horses only.

'Grave doubts existed as to the practicability of working a large traffic by means of travelling engines.

'The directors were inundated with schemes of all sorts for facilitating locomotion. The projectors of England, France, and America, seemed to be let loose upon them. There were plans for working the waggons along the line by water-power. Some proposed hydrogen, and others carbonic acid gas. Atmospheric pressure had its eager advocates. And various kinds of fixed and locomotive steam-power were suggested. The directors felt themselves quite unable to choose from amidst this multitude of projects. (Mr. Stephenson) expressed himself as decidedly as heretofore in favour of smooth rails and locomotive engines, which, he was confident, would be found the most economical and by far the most convenient moving power that could be employed. The Stockton and Darlington Railway being now at work, another deputation went down personally to inspect the fixed and locomotive engines on that line, as well as at Hetton and Killingworth. They returned to Liverpool with much information; but their testimony as to the relative merits of the two kinds of engines was so contradictory, that the directors were as far from a decision as ever.

'Such was the result, so far, of George Stephenson's labours. He had scarcely an adherent, and the locomotive system seemed on the eve of being abandoned. Still he did not despair. With the profession as well as public opinion against him—for the most frightful stories were abroad respecting the dangers ... which the locomotive would create—Stephenson held to his purpose. Even in this, apparently the darkest hour of the locomotive, he did not hesitate to declare that locomotive railroads would, before many years had passed, be "the great highways of the world".

'He urged his views upon the directors in all ways, and, as some of them thought, at all seasons. At length, influenced by his persistent earnestness not less than by his arguments, the directors determined to offer a prize of £500 for the best locomotive engine, which, on a certain day, should be produced on the railway, and perform certain specified conditions in the most satisfactory manner.

'The conditions were these:

1. The engine must effectually consume its own smoke.
2. The engine, if of six tons weight, must be able to draw after it, day by day, twenty tons weight (including the tender and water-tank) at *ten miles* an hour, with a pressure of steam on the boiler not exceeding fifty pounds to the square inch.

3. The boiler must have two safety-valves, neither of which must be fastened down, and one of them be completely out of the control of the engineman.

4. The engine and boiler must be supported on springs, and rest on six wheels, the height of the whole not exceeding fifteen feet to the top of the chimney.

5. The engine, with water, must not weigh more than six tons; but an engine of less weight would be preferred on its drawing a proportionate load behind it; if only four and a half tons, then it might be put on only four wheels. The Company to be at liberty to test the boiler, etc., by a pressure of one hundred and fifty pounds to the square inch.

6. A mercurial gauge must be affixed to the machine, showing the steam pressure above forty-five pounds per square inch.

7. The engine must be delivered, complete and ready for trial, at the Liverpool end of the railway, not later than 1 October 1829.

8. The price of the engine must not exceed £550.'

The locomotive was the least of George Stephenson's problems, and he left the actual construction in the hands of his son Robert and of Mr. Henry Booth, the Secretary of the Railway Company.

'On the day appointed for the great competition of locomotives at Rainhill, the following engines were entered for the prize:

1. Messrs. Braithwaite and Ericsson's "Novelty".
2. Mr. Timothy Hackworth's "Sanspareil".
3. Messrs. R. Stephenson and Co.'s "Rocket".
4. Mr. Burstall's "Perseverance".

'Another engine was entered by Mr. Brandreth of Liverpool—the "Cycloped", weighing 3 tons, worked by a horse in a frame, but it could not be admitted to the competition. The above were the only four exhibited, out of a considerable number of engines constructed in different parts of the country in anticipation of this contest, many of which could not be satisfactorily completed by the day of trial.

'The ground on which the engines were to be tried was a level piece of railroad, about two miles in length. Each was required to make twenty trips, or equal to a journey of 70 miles, in the course of the day; and the average rate of travelling was to be not under 10 miles an hour. It was determined that, to avoid confusion, each engine should be tried separately, and on different days.

'The day fixed for the competition was 1 October (1829), but to allow sufficient time to get the locomotives into good working order, the directors extended it to the 6th. On the morning of the 6th, the ground at Rainhill presented a lively appearance, and there was as much excitement as if the St. Leger were about to be run. Many thousand spectators looked on, among whom were some of the first engineers and mechanicians of the day. A stand was provided for the ladies; the "beauty and fashion" of the neighbourhood were present, and the side of the railroad was lined with carriages of all descriptions:

'It was quite characteristic of the Stephensons, that, although their engine did not stand first on the list for trial, it was the first that was ready; and it was accordingly ordered out by the judges for an experimental trip, which it made quite successfully. It ran about 12 miles, without interruption, in about 53 minutes.

'The "Novelty" was next called out. It was a light engine, very compact in appearance, carrying the water and fuel upon the same wheels as the engine. The weight of the whole was only 3 tons and 1 hundredweight. A peculiarity of this engine was that the air was driven or *forced* through the fire by means of bellows. The day being now far advanced, and some dispute having arisen as to the method of assigning the proper load for the "Novelty", no particular experiment was made, further than that the engine traversed the line by way of exhibition, occasionally moving at the rate of 24 miles an hour. The "Sanspareil", constructed by Mr. Timothy Hackworth, was next exhibited; but no particular experiment was made with it on this day.

'The contest was postponed until the following day, but before the judges arrived on the ground, the bellows for creating the blast in the "Novelty" gave way, and it was found incapable of going through its performance. A defect was also detected in the boiler of the "Sanspareil"; and some further time was allowed to get it repaired. The large number of spectators who had assembled to witness the contest were greatly disappointed at this postponement; but, to lessen it, Stephenson again brought out the "Rocket", and, attaching to it a coach containing thirty persons, he ran them along the line at the rate of from 24 to 30 miles an hour, much to their gratification and amazement. Before separating, the judges ordered the engine to be in readiness by eight o'clock on the following morning, to go through its definitive trial according to the prescribed conditions.

'On the morning of 8 October, the "Rocket" was again ready for the contest. The engine was taken to the extremity of the stage, the firebox was filled with coke, the fire lighted, and the steam raised until it lifted the safety-valve loaded to a pressure of 50 pounds to the square inch. This proceeding occupied fifty-seven minutes. The engine then started on its journey, dragging after it about 13 tons weight in waggons, and made the first ten trips backwards and forwards along the two miles of road, running the 35 miles, including stoppages, in one hour and 48 minutes. The second ten trips were in like manner performed in 2 hours and 3 minutes. The maximum velocity attained during the trial trip was 29 miles an hour, or about three times the speed that one of the judges of the competition had declared to be the limit of possibility. The average speed at which the whole of the journeys were performed was 15 miles an hour, or 5 miles beyond the rate specified in the conditions published by the Company. The entire performance excited the greatest astonishment among the assembled spectators; the directors felt confident that their enterprise was now on the eve of success; and George Stephenson rejoiced to think that in spite of all false prophets and fickle counsellors, the locomotive system was now safe. When the "Rocket", having performed all the conditions of the contest, arrived at the "grandstand" at the close of its day's successful run, Mr. Cropper—one of the directors initially favourable to the fixed-engine system—lifted up his hands, and exclaimed, "Now has George Stephenson at last delivered himself!"

'The "Sanspareil" was not ready until the 13th; and when its boiler and tender were filled with water, it was found to weigh 4 cwt. beyond the weight specified in the published conditions as the limit of four-wheeled engines; nevertheless the judges allowed it to run on the same footing as the other engines, to enable them to ascertain whether its merits entitled it to favourable consideration. It travelled at the average speed of about 14 miles an hour, with its load attached; but at the eighth trip the cold-water pump got wrong, and the engine could proceed no further.

'The "Rocket" was thus the only engine that had performed, and more than performed, all the stipulated conditions; and its owners were declared to be fully entitled to the prize of £500, which was awarded to the Messrs. Stephenson and Booth accordingly. And further, to show that the engine had been working quite within its powers, Mr. Stephenson ordered it to be brought upon the ground and detached from all incumbrances, when, in making two trips, it was found to travel at the astonishing rate of 35 miles an hour.'

We all know that George Stephenson was a hero in his own generation. So too will be the men who make the first commercial breeder reactor in our own generation. But let us follow George Stephenson to his final days:

'At home he lived the life of a country gentleman, enjoying his garden and grounds, and indulging his love of nature, which, through all his busy life, had never left him. It was not until the year 1845 that he took an active interest in horticultural pursuits. Then he began to build new melon-houses, pineries, and vineries, of great extent; and he now seemed as eager to excel all other growers of exotic plants in his neighbourhood, as he had been to surpass the villagers of Killingworth in the production of gigantic cabbages and cauliflowers some thirty years before. He had a pine-house built 68 feet in length and a pinery 140 feet. Workmen were constantly employed in enlarging them, until at length he had no fewer than ten glass forcing-houses, heated with hot water, which he was one of the first in that neighbourhood to make use of for such a purpose. He did not take so much pleasure in flowers as in fruits. At one of the county agricultural meetings, he said that he intended yet to grow pine-apples as big as pumpkins. The only man to whom he would "knock under" was his friend Paxton, the gardener to the Duke of Devonshire; and he was so old in the service, and so skilful, that he could scarcely hope to beat him. Yet his "Queen" pines did take first prize at a competition with the Duke, though this was not until shortly after his death, when the plants had become more fully grown. His grapes also took the first prize at Rotherham, at a competition open to all England. He was extremely successful in producing melons, having invented a method of suspending them in baskets of wire gauze, which, by relieving the stalk from tension, allowed nutrition to proceed more freely, and better enabled the fruit to grow and ripen.

'He took much pride also in his growth of cucumbers. He raised them very fine and large, but he could not make them grow straight. Place them as he would, notwithstanding all his propping of them, and humouring them by modifying the application of heat and the admission of light for the purpose of effecting his object, they would still insist on growing crooked in their own way. At last he had a number of glass cylinders made at Newcastle, for the purpose of an experiment; into these the growing cucumbers were inserted, and then he succeeded in growing them perfectly straight. Carrying one of the new products into his house one day, and exhibiting it to a party of visitors, he told them of the expedient he had adopted, and added gleefully, "I think I have bothered them noo!"'

Index